Natural Gas:
an International Perspective

Natural Gas: an International Perspective

Proceedings of the Oxford Energy Seminar
1982–1985

EDITED BY
ROBERT MABRO

Published by the Oxford University Press
for the Oxford Institute for Energy Studies
1986

Oxford University Press, Walton Street, Oxford OX2 6DP
Oxford New York Toronto
Delhi Bombay Calcutta Madras Karachi
Petaling Jaya Singapore Hong Kong Tokyo
Nairobi Dar es Salaam Cape Town
Melbourne Auckland
and associated companies in
Beirut Berlin Ibadan Nicosia

British Library Cataloguing in Publication Data
Natural gas: an international perspective: proceedings of the Oxford Energy Seminar 1982–1985.
1. Gas, Natural
I. Mabro, Robert II. Oxford Institute for Energy Studies
333.8'233 HD9581.A2
ISBN 0-19-730002-2

Typeset by Joshua Associates Limited, Oxford
Printed in Great Britain by
Billing and Sons Ltd., Worcester

CONTRIBUTORS

Dr Burckhard Bergmann	Executive Board, Ruhrgas AG, West Germany
Mr James T. Jensen	President, Jensen Associates, USA
Mr Arve Johnsen	President, Statoil, Norway
Mr Aman R. Khan	President, GDC Inc, USA
Mr Robert Mabro	Director, Oxford Institute for Energy Studies, UK
Mr Ahmed Messili	Head, Department of Information and Statistics, Sonatrach, Algeria
Mr Jonathan P. Stern	Head of Joint Energy Programme, Royal Institute of International Affairs and Policy Studies Institute
Mr Ir Wijarso	Adviser to the Minister of Mines and Energy, Indonesia

The *Oxford Energy Seminar* is an international conference designed for government officials, industrialists, managers and other professionals engaged in the field of energy. The Seminar is sponsored by St Catherine's College and co-sponsored by OPEC and OAPEC. It was established in 1979 and its sessions are held annually at St Catherine's over a two-week period in September.

The Seminar, which is fully residential, has educational objectives. Its aims are to enhance the professional qualifications of participants; to improve the understanding of factors and forces which influence the operations of energy markets and the behaviour of policy-makers and economic agents; and to provide a privileged opportunity for contacts and debate between participants from petroleum exporting and importing countries.

A most distinctive feature of the Seminar is the balance of speakers and participants between nationals of petroleum exporting and importing countries and between representatives of government and industry. Each annual session of the Seminar involves fifty to sixty participants and approximately forty speakers. The organization is delegated by the co-sponsors to a Board of Management consisting of the Secretary General of OPEC, the Secretary General of OAPEC, Mr Wilfrid Knapp and Mr Robert Mabro.

The *Oxford Institute for Energy Studies* was established in 1982 to pursue through research and advanced study of problems of international energy the same objectives as the Seminar. The Institute is an association of members, including Oxford University and three of its constituent colleges, OAPEC, the EEC, the UK Department of Energy and other national or regional institutions from Sweden, Norway, France, Japan, Mexico and the Arab world.

The Institute is committed to the idea of co-operation between scholars representing different sides of the international energy debate. This international character is reflected in the composition of both the membership and the research staff of the Institute. The Institute is also committed to achieving high academic standards in the study of real-life problems in the energy world. The aim is to combine excellence in research and relevance to important policy issues.

The Institute has a board of governors representing all members. Dr Ali Attiga is the Chairman, Sir Rex Richards the Vice-Chairman and Robert Mabro the Managing Director.

The views and opinions expressed in this book are those of the respective authors. They do not necessarily reflect those of the governments, companies or organizations to which they belong. They should not be construed as representing the views of the Oxford Energy Seminar, the Oxford Institute for Energy Studies, the sponsors, members or management of these two institutions.

ACKNOWLEDGEMENTS

The co-operation of Judy Mabro who edited the original typescripts for consistency and clarity, of David Guthrie who followed the book through all production stages, and of Margaret Ko who never tired of typing and endless corrections, is most gratefully acknowledged. Jonathan Stern provided much valued editiorial advice. Thanks are due to Karen Exley and many others who contributed every year to the efficient running of the Seminar and to David Murray who designed the cover of the book.

CONTENTS

Introduction 1

1. The Prospects for International Trade in Natural Gas 9
Robert Mabro

2. The Encroachment of Natural Gas in the Oil Market 23
Aman R. Khan

3. Natural Gas in Western Europe: Modest Growth in a Buyers' Market? 43
Burckhard Bergmann

4. Norwegian Gas: Reserves, Production and Markets 71
Arve Johnsen

5. The Prospects for Gas in Developing Countries 81
Aman R. Khan

6. The Evolution of Market-responsive Pricing for Natural Gas in the USA and its Effect on Gas Trade 95
James T. Jensen

7. International Gas Trade and Respect of Contracts 111
Ahmed Messili

8. Present and Possible Future Utilisation of Natural Gas in the Western Pacific Region 125
Ir Wijarso

9. Government Intervention as a Factor in International Gas Trade: Help or Hindrance? 133
Jonathan P. Stern

10. Natural Gas Development: Pricing, Politics and Profitability 141
Aman R. Khan

TABLES

2.1 Energy Costs to Consumer Including Taxes. 30
2.2 World Gas Demand, by Region. 33
2.3 Gas Demand, by Market Sector. 34
2.4 Outlook for Gas Demand by Sectors in the USA. 37

3.1 Final Energy Consumption in Western Europe, by Market Sector. 1979–84. 46
3.2 Percentage Shares of Fuels in Energy Consumption of West European Countries. 1983. 47
3.3 Natural Gas Consumption in Western Europe, by Market Sector. 1979–84. 48
3.4 Energy Consumption. 1982 and 1984. 50
3.5 Energy Consumption in West Germany. Percentage Changes from 1973 to 1985. 51
3.6 Trends in Supply and Demand of Natural Gas in Western Europe. 62
3.7 Soviet Natural Gas Production and Exports. 65

5.1 World Natural Gas Reserves and Commercial Production. 83
5.2 Classification of Developing Countries with Natural Gas Resources by Region, Income Group and Net Trade in Oil. 84
5.3 Gas Reserves and Production in Selected Developing Countries. 85
5.4 Possible Uses of Gas by Industries. 88
5.5 Natural Gas Requirements for Typical Industrial Usage. 89
5.6 Selected Developing Countries: Comparative Costs of Natural Gas Production. 90

8.1 Imports of LNG to Japan, by Country. 1984. 127

FIGURES

2.1	Relation of Non-Communist World Crude Oil Consumption to OPEC Marker Crude Oil Price.	25
2.2	Residual Oil Consumption in the Non-Communist World.	27
2.3	Distillate Oil Consumption in the Non-Communist World.	28
3.1	The Transformation of the West European Energy Market. 1960–84.	44
3.2	Changes in Western Europe since 1979.	45
3.3	Price and Consumption Indices for Various Sources of Energy in West Germany since 1978.	48
3.4	Natural Gas Market Penetration in Western Europe, by Sector.	49
3.5	Primary Energy Demand Predictions for West Germany.	53
3.6	Historic and Preducted Trends of Economic Growth in West Germany, the USA and Japan.	54
3.7	The Share of Natural Gas in World Energy Markets.	59
3.8	Natural Gas Supplies to Western Europe.	61
3.9	Norwegian North Sea Gas Supplies to Western Europe.	64
4.1	Norwegian Gas Provinces and Transportation System.	72
10.1	Selected LNG Prices. 1972–84.	152

ABBREVIATIONS

AG	Aktiengesellschaft (Joint-stock company)
ASEAN	Association of South East Asian Nations
bcf	billion cubic feet
bcf/d	billion cubic feet per day
bcm	billion cubic metres
bcm/d	billion cubic metres per day
b/d	barrels per day
boe	barrels of oil equivalent
boe/d	barrels of oil equivalent per day
BP	British Petroleum
btoe	billion tonnes of oil equivalent
Btu	British thermal units
c.i.f.	cost, insurance and freight
CNG	Compressed natural gas
CPE	Centrally planned economy
EEC	European Economic Community
ERA	Energy Regulatory Administration (USA)
FERC	Federal Energy Regulatory Commission (USA)
f.o.b.	free on board
FPC	Federal Power Commission (USA)
GDP	Gross domestic product
GNP	Gross national product
IEA	International Energy Agency (OECD)
IGU	International Gas Union
km	kilometres
kWh	kilowatt hours
LNG	Liquefied natural gas
LPG	Liquefied petroleum gas
mb/d	million barrels per day
mboe/d	million barrels of oil equivalent per day
mBtu	million British thermal units
mcf	million cubic feet
mcf/d	million cubic feet per day
MITI	Ministry of International Trade and Industry (Japan)
mtoe	million tonnes of oil equivalent

MW	megawatts
n.a.	not available
NGLs	Natural gas liquids
NGPA	Natural Gas Policy Act
OECD	Organisation for Economic Co-operation and Development
OIDCs	Oil-importing developing countries
OPEC	Organization of the Petroleum Exporting Countries
PEC	Primary energy consumption
ROR	Rate of return
scf	standard cubic feet
SMP	Special marketing programme
tcf	trillion cubic feet
UAE	United Arab Emirates
UKCS	United Kingdom Continental Shelf

INTRODUCTION

Robert Mabro

This volume is part of a series of books and monographs on the economics and politics of energy. The series is sponsored by the Oxford Institute for Energy Studies and published by the Oxford University Press. Its purpose is to disseminate to a wide audience research work undertaken by the Institute and the most important papers presented at the annual sessions of the Oxford Energy Seminar. The series will also include bibliographies, statistical material and other reference books relevant to the advanced study of energy. The aim is to reach industrialists, politicians, government officials, students, consultants and academics, and those members of the public with an interest in issues of international importance.

Natural Gas: an International Perspective is a collection of papers written by well-known specialists on a fuel which was often neglected in the past but which is acquiring now an increasingly significant role. The authors do not all belong to the same side of the energy debate, and their contributions naturally involve different positions and points of view. The philosophy of the Oxford Energy Seminar, the forum in which most of these papers were presented,[1] is to promote a debate between energy consumers and energy producers, between government and industry, and between the analysts and the practitioners, particularly those responsible for policy and for major economic decisions.

The reader should therefore not be surprised to find that different authors contradict each other on certain issues, not only as regards ideas and conclusions, but sometimes in the presentation and interpretation of the same set of facts.

[1] These are Bergmann's paper on Western Europe, Johnsen's on Norway, Khan's three contributions, the chapter by Messili on the respect of contracts and Wijarso's paper on the Western Pacific.

Normally the Editor's job is to remove contradictions and reconcile divergences. In this book, as in the Seminar, differences in opinion are allowed to manifest themselves as clearly as possible wherever they arise. Editorial interference is kept to the minimum and the readers are given the opportunity to assess for themselves the different sides of the debate.

To allow different views to emerge within the pages of a book does not preclude the need to organize the work in a coherent and well-structured manner. An anthology of Seminar papers may involve serious gaps either because certain topics were not presented at all or because their treatment did not give rise to a publishable paper. For this reason we have enlarged the collection by including three papers (Jensen's, Mabro's and Stern's) written by authors who are associated in some ways with the work of the Seminar or the Institute.[2] Although the book does not attempt to cover comprehensively all the aspects of the gas issue, the inclusion of these additional papers was necessary to improve both its structure and its balance.

Natural gas accounts for a large share of world energy consumption but does not occupy a commensurate place in international energy trade. Furthermore the international market for gas is fragmented. The concept of a 'world gas market' can be misleading, and it is more appropriate to distinguish separate sets of bilateral or multilateral trade relationships. In today's world, the four main sets, identified for convenience by the destination of natural gas exports, are Western Europe (108.2 bcm in 1984), Japan (35.7 bcm), Eastern Europe (32.8 bcm) and the USA (23.9 bcm).[3] The papers presented in this book cover three of these sets: the one left out is Eastern Europe where all the countries of the region as defined here import gas exclusively from the USSR. Naturally Western Europe receives more coverage than the USA and Japan, partly because the volume of gas trade is much larger, and partly because the problems involved are sometimes more complex.

[2] Jensen is a regular speaker at the Seminar and Stern a valued academic referee of work done at the Institute.

[3] *BP Review of World Gas*, 1985, Table 12. In Eastern Europe imports of gas we have excluded the USSR from the definition of this region and included Finland and Yugoslavia.

However the book does not focus on regions only. The main concern is with issues, and this also applies to papers which refer to a country or a geographical area in their titles. The first issue is the prospects for international trade in natural gas. This is the main subject matter of Chapter 1, but most authors address it at some length in their own papers. Indeed one of the objectives of the book is to assess the constraints which inhibit trade and the opportunities for expansion which may lie ahead.

The main obstacles to trade are the large capital costs of gas supply systems, and the high level of risk associated with the long lead-in times of investment and the rigidities of both pipeline and LNG chains. These technical rigidities involve the exporter and the importer of natural gas in long-term contracts which may be difficult to enforce throughout their stipulated life and which may involve unforeseen losses for the parties when future market circumstances begin to diverge in a significant way from those envisaged at the time of the initial agreement. Considerable uncertainty surrounds future developments of energy prices, alternative supplies of competing fuels and energy demand.

That trade relationships can turn sour and contracts fail to provide the security which they are intended to provide is abundantly illustrated by the breakdown of the relationship between Algeria and its US partners. In this book the Algerian position is clearly expressed by Messili in Chapter 7. It is important to lend an attentive ear to the Algerian argument whether one agrees with it or not. The first reason is that Algeria has a considerable production and export potential in natural gas, considering the size of its reserves, the LNG investments already made and its proximity to the southern shores of Europe. The realization of this potential is of interest both to Algeria and to those who constitute its natural markets; and the prerequisite for such developments is a sympathetic dialogue which involves a careful understanding of the partners' positions.

The second reason is that the Algerian line relates to a set of Third World aspirations which still have considerable political force. The beliefs, (a) that natural resources are a unique source of national wealth which should not be squandered in response to the short-term signals of erratic markets, and (b) that the logic of *laissez-faire*, unregulated competition and self-interested private enterprise is not necessarily the logic which serves best the long-term interests of developing countries handicapped by a lack of bargaining power, may have lost some ground in developing countries. They continue however to appeal and they are likely to re-emerge in the future.

The Algerian position on gas may not have helped the immediate economic interests of that country, but they are a powerful reminder of the fundamental divergences between the objectives of developing and rich industrial countries. Long-term gas agreements which compromise the objectives of the Third World exporters, and which are often hailed as a vindication of commercial realism, may well lack the solid foundations which guarantee their permanence.

The role of political factors in determining the behaviour of economic agents in international gas trade is not restricted to the exporting nations of the Third World. Stern's paper (Chapter 9) suggests that the governments of industrialized countries are increasingly intervening in this area and that their interference is both highly visible and significant. He identifies five main types of intervention ranging from the blunt refusal to allow a domestic utility to sign a contract agreed on a commercial basis with a foreign exporter to the more subtle official support given to suppliers of heavy equipment and technology when an international gas contract involves large investments in pipelines or LNG facilities.

All these political considerations, critical as they are, do not contribute the whole of the story. The development of gas production and trade is primarily an economic issue involving such factors as demand, supply costs, investments and prices. Politics may interfere at all levels and help or hinder the final outcome but its exact role cannot be assessed without a correct understanding of the economics of the gas problem.

The first economic issue worth considering is demand. Several papers in this collection attempt to assess the demand prospects for gas in the three major industrialized areas, the USA, Western Europe and Japan, and in developing countries. Bergmann (Chapter 3) examines the issue extensively for Western Europe; Mabro (Chapter 1), Khan (Chaper 2) and others discuss demand prospects in Europe and other areas. It is evident that the expansion of gas demand is a function of both economic growth and inter-fuel substitution (relative prices). Khan argues that gas encroached on the petroleum market during the late 1970s and early 1980s because OPEC's actions pushed up the price of oil to uneconomically high levels; and that gas would have gained an even larger share of world energy consumption if its own price was not artificially pegged to oil in many countries. Looking ahead towards the year 2000, Bergmann suggests that gas demand in Western Europe will not increase very much, largely because of low economic growth forecast, and partly because of pricing problems.

Energy policy-makers and planners cannot influence very much the rates of world economic growth; they can only take them as exogenous variables. However they can take a view about gas pricing and choose the appropriate regime. An interesting question is whether gas prices should be entirely left to the operations of free-market – i.e. perfectly competitive – forces? In his three contributions Khan emphasises the merits of unregulated markets and free consumer choice. Bergmann, who often refers in his paper to the determinant role played by politics in important energy decisions in Western Europe (reminding us, like Stern, that political meddling is not exclusive to Third World countries), also insists that competitive pricing is essential. 'It is therefore necessary to agree a pricing mechanism which guarantees that gas will always be competitively priced for the end-user. . . .'[4]

Yet Bergmann is also concerned about the reliability of supplies. As mentioned earlier gas supply systems are capital-intensive and fairly rigid. Disruptions can be extremely costly. He recognises that 'the long-term nature of the [gas] investment must be matched by an equivalent period of amortisation.'[5] In other words the investors may legitimately seek prices which yield over a long period of time the necessary return. This is where the issue becomes thorny. Competitive pricing is required to secure a market but it is not difficult to imagine circumstances when it will fail to cover the total costs of supply.

From a different perspective, that of a producer and an

[4] See below p. 68. [5] See below p. 68.

exporter, Johnsen (Chapter 4) suggests, albeit indirectly, that supplies from a stable and secure area such as Norway justify to some extent the development of high-cost reserves. 'It must be emphasised that Norwegian gas will never be a cheap gas, partly because of high costs and partly because deliveries are based on long-term agreements from a *stable* area' (our italics).[6] His paper, written before the breakdown of the Statoil/British Gas preliminary agreement over Sleipner, is an interesting document on the Norwegian position during the negotiations with the UK. Was Norway bargaining about a security of supply premium, or was its position determined by a perception of future UK dependence on Norway's gas supplies?

The authors who argue from a consumer viewpoint in favour of competitive pricing and those who insist as producers on including an additional element, a hidden or an explicit premium, fail to inform us of the great complexities of the price issue. In fact this is not their aim. Their conflicting messages are simple but powerful. On the one hand Khan tells us that producers can only ignore the verdict of markets at their peril. On the other hand, Johnsen and Messili are saying that competitive prices, in certain instances, may be too low to elicit the appropriate supply responses. High-cost gas reserves will not be developed; conservationists may not wish to deplete their exhaustible resources at low prices; and supply security which is essential to all parties can be seriously threatened by low and volatile prices.

It is as difficult to reject outright either message as it is to reconcile them. Taken together they simply reveal a fundamental mismatch between supply and demand conditions for gas, a problem which goes a long way towards explaining why gas remains oil's neglected brother.

Expressed differently, the problem of gas development is that gas has to compete with oil. As indicated in Chapter 1, the price of oil includes a much higher rent per unit than the price of gas. The difference was very striking in 1984 when the paper was written, and remains so now in 1986 despite the fall in petroleum prices. The existence of a large rental element in oil, which producers can choose to forgo in a price war (or which

[6] See below p. 79.

the market can erode), threatens the gas investor. He is inevitably inhibited by considerable uncertainty. Although he could attempt to increase the demand for his fuel by competing against a high oil price, any sharp downward movement of this price would destroy the profitability of a very costly, long-term investment. Paradoxically the future of the gas market depends more on the expected behaviour of oil prices than on the pricing policies of gas suppliers.

In the final analysis the key to gas development is economic growth leading to an expansion of energy demand. In the absence of growth, gas can only increase its market share through inter-fuel substitution; and the room for advances on this front, however significant in the initial stages, is ultimately limited. Growth can set a floor to oil prices – a necessary protection for investments in gas. In a growth context gas and other fuels will be able to compete with oil and set a ceiling to its price, thus preventing unreasonable rises with damaging effects on energy demand. Under certain conditions, economic growth can provide a way out.

The energy sector and the world economy have been in a state of stagnation and crisis in the past ten or twelve years. The ills of the economy have affected energy and vice versa. Economic stagnation was blamed on oil and was probably made worse by policies designed to cure other ills. Like many other industries gas also suffered from price distortions and constraints on demand. Growth can initiate a virtuous circle but economic expansion requires careful management of the domestic economy and a good measure of international co-operation. In the energy field the development of competing supplies which can restore efficient price relationships between fuels, as suggested above, involves time lags and long phases. Markets left to their own devices may fail to induce smooth transitions. There is a case after all for intervention by governments of a beneficial type and of co-operation between consuming and producing countries.

1 THE PROSPECTS FOR INTERNATIONAL TRADE IN NATURAL GAS*

Robert Mabro

Natural gas is a major energy source with attractive properties. In 1984 it supplied about 20 per cent of world primary energy consumption and ranked third in the fuel list after oil and coal. However, the share of natural gas in energy consumption does not reflect correctly the importance of known reserves. The *BP Review of World Gas* estimates gas reserves (the published proved concept) at 85.6 btoe and oil reserves at 96.1 billion tonnes.[1] According to these estimates, there is virtually as much gas as oil in the world although in reality there may be more gas because exploration so far has been biased in favour of oil. Gas discoveries were often the by-product of oil exploration.

Natural gas is not as yet a much traded commodity in international commerce. In 1984 total gas exports (both by pipeline and as LNG) amounted to 12.5 per cent of gas production, while in the same year oil exports represented over 40 per cent of world oil output. Since world oil production is much larger than world gas output (measured in equivalent thermic units), the volume of internationally traded gas appears to be small compared with oil.

The purpose of this paper is to assess the prospects for international trade in gas, and in order to do so we need to ask some preliminary questions:

(a) Why is the share of gas in total energy small compared with that of oil?
(b) Why is world trade in gas so limited compared to the volume of production?

* Paper presented at the Eighth Doha Technical Symposium, Qatar, November 1984.

[1] *BP Review of World Gas*, 1985.

An answer to these questions will give us some indication of the constraints which may limit gas use and gas trade. Having identified the constraints we shall then consider the opportunities. A brief review of the three main markets, the USA, Western Europe and Japan, will reveal the possible imbalances between domestic supply and demand which may open up an opportunity for trade. Although economic growth and relative prices are the key to future prospects, this paper will also consider the need to improve contractual arrangements and to devote research to the complex issue of how to design 'good' contracts.

Oil and gas belong to the same fuel family, and are often found together in the same geological formation or indeed in the same reservoirs (hence, the term associated gas). In the past gas was considered to be a valueless by-product of oil production and was flared in large quantities *in situ*. It was therefore dubbed 'oil's little brother', for it was certainly a neglected member of the family.[2]

Yet gas is not oil and as James Jensen emphasised in an earlier paper, we must forget the similarities and focus on the differences between the *properties* of these two fuels if we want to understand the differences in the *behaviour* of demand and in the *patterns* of trade.[3] The physical properties of gas (compared with those of oil) do not fully explain its status as 'oil's little brother'. Gas is a clean fuel with very little emission when it burns. It can be carried in bulk (say, by pipe) or in small amounts (say, in metal bottles) to the place of use. The intensity and direction of its flame can be quickly adjusted to suit different purposes.

These properties provide gas with an advantage over oil in domestic and industrial uses. Gas is a good feedstock for the petrochemical industry: in some instances gas is a specific input, and in many others it may be substituted for another feedstock. Like all fuels, gas can be used to raise steam, and can therefore compete with heavy oil and coal in the public utilities sector. Finally, gas is a superior fuel from an environmental point of view. However, there are problems of safety associated

[2] Aït-Laoussine, N., 'Gas: Recent Developments and Problems of Supply', Mabro (ed), *World Energy: Issues and Policies*, OUP, 1980.

[3] Jensen, J., 'World Natural Gas Reserves and the Potential for Gas Trade', Mabro (ed), *op.cit.*

with gas, which is liable to provoke explosions. These problems are important in offshore operations, where a blow-out could endanger life, and in distribution, especially in the urban centres of poor countries, where maintenance leaves much to be desired and where the rigorous application of safety measures cannot be taken for granted.

Bearing in mind this last reservation, we have listed many characteristics that make gas an attractive source of energy. In so far as its 'intrinsic fuel' properties are concerned there is no impediment to an expansion of gas consumption in certain energy sectors, although in others, such as transport, a serious limitation does exist. The energy-using capital stocks now extant in transport are largely designed to consume oil products. Some cars can run on LPG but they constitute a very small proportion of the automobile stock. In short, gas may be more attractive than oil *at the point of use* of the fuel in certain sectors, but its penetration of the transport sector is severely restricted by a formidable technological barrier. Oil remains the most versatile fuel as there is virtually no energy use it cannot meet. The same cannot be said of either gas or coal.

If 'intrinsic fuel' properties do not wholly explain the low share of gas in energy demand and trade we should seek other explanatory factors. Let us turn first to other natural or geo-physical features. The main difference between gas and oil is the high cost of gathering and transporting gas, whereas the oil trade is helped by low-cost tanker transportation. Some analysts maintain that trade in oil was made truly international by the flexibility and cheapness of the transport system, for a tanker can load in any export terminal and carry its cargo to any major harbour in the world. Oil can be transported in ultra-large vessels, thus enjoying the benefits of economies of scale, or in very small ships, thus enjoying the benefits of flexible access to small and big ports alike. A crude tanker is a fairly rudimentary 'floating container' with little specialized equipment. The loading/unloading of crude oil requires certain specific installations, but they are neither technologically complex nor financially prohibitive.

Crude oil moves in bulk by pipe from the field to the export terminal and by tanker from terminals to anywhere in the world. Gas moves by pipe when transported in its original form between (or within) countries or by LNG tankers. But sea transport by vessels requires liquefaction at the point of shipment and regasification at the point of delivery. The necessary installations at these two points are considerable and expensive; the LNG tankers are elaborate vessels with specific equipment; and the LNG transport line is inevitably rigid. In most cases the LNG tankers can only link a given terminal with a given import point.

According to Jensen, 'gas pipeline costs (per Btu) are generally higher than oil pipeline costs and the relative disadvantage is greater for small scale movements'.[4] There are problems of scale, and the transport of gas from a small discovery may therefore prove impractical. A small-volume pipeline over a relatively short distance across hostile sea may prove more expensive (per unit) than a big-volume LNG transport over a long distance. In short, gas transport involves significant rigidities and substantial cost penalties against small fields. These features of gas transport constitute a serious impediment to the development of international trade.

The implications of this state of affairs are numerous. First, the rigidity of gas supply systems (whether the transportation vehicle is a pipeline or a ship) involves buyers and sellers in specific relationships. If exports from source A are rigidly tied by the supply system to an input outlet in market B, then A and B are joined together in a bilateral deal with no option for transferring custom in the short or medium term. In other words A cannot immediately divert its exports to another customer when some disagreement arises with B. Diversion of trade requires investment in a new project involving large sums of money and long periods of time. B may also find that he cannot replace at short notice the gas obtained from A, unless he has diverse access to energy imports and some redundancy in his alternative supply systems.

Secondly, the rigidity of gas supply systems leads to a fragmentation of the international gas trade. One finds a number of bilateral relationships rather than multilateral trade, and this characteristic is accentuated by the impact of distance on transport costs. Since transport cost is a very

[4] *ibid.*

large proportion of the price of delivered gas (in international trade) a point is soon reached beyond which it is uneconomical to move gas. This feature, of course, is common to all traded commodities but its significance varies from one commodity to another depending on the nature of transport costs. In oil trade distance imposes a much smaller penalty than in gas.

Thirdly, gas supply systems are capital intensive. In addition, the lead-in times are long, and the rigidity of the systems involves a large element of risk since the viability of projects partly depends on the two parties (buyer and seller) abiding by the contract over a long period of time. In other words, the financing of large gas projects raises the issue of supply/demand security.

The limitations on international gas trade, then, arise from high transport costs, high capital costs and rigid bilateral trading relationships arising from the technical characteristics of supply systems. Thus it is difficult to design *enforceable* long-term contracts that protect the interests of the producer, the importer and the financier. The producer needs a secure market, the buyer a secure supply at an agreed price and the financier guaranteed repayment of the interest and principal of his loans. However, no producer can in fact commit himself to a price that will satisfy the buyer over a period of fifteen or twenty years, and no buyer can be sure that his demand for gas will correspond throughout the contract period to the contracted volumes. Furthermore, no financier can be certain that his clients (the producer and the customer) will avoid a disastrous breakdown of their trading relationship – a breakdown that may impoverish them to the point of default.

Having identified the constraints on gas trade, let us now turn to the opportunities. The first step is to survey the characteristics of the major gas markets.

The *USA* is by far the world's largest natural gas market. Substantial reserves and ease of transport by pipelines over land enabled the US gas industry to move this fuel from producing areas, mainly in the southern parts of the country, to consumption centres throughout the nation. US consumption was estimated at 514.6 bcm in 1984, in decline from a peak of

660.0 bcm attained in 1972.[5] The sectoral composition of US gas demand in 1983 was as follows:

Sector	%
Domestic/commercial	43
Industrial	37
Public utilities	20
Total	100

Although the US gas market is almost self-sufficient, its sheer size can provide opportunities for international trade through changes in the domestic supply/demand balance. A small percentage gap in this balance can amount to a significant volume in absolute terms. At present the US gas situation seems to suffer from an imbalance involving excess supplies, which the industry describes in terms of a gas bubble. As mentioned earlier, consumption has declined in the past decade. The causes are many:

(a) Regulation of fuel uses which pushed gas out of electric power generation (this factor was important in the 1970s but ceased to be significant after 1981).
(b) The economic recession.
(c) Past rises in the price of natural gas in the domestic US market (the well-head price rose by 760 per cent between 1973 and 1981), which induced significant conservation use, even though gas prices remain lower than the prices of competing oil products.

The fall in demand may be accompanied by a temporary increase in availability, because the price rises resulting from deregulation (the Natural Gas Policy Act of 1978) could induce higher drilling rates. The problem is that the reserve base is not sufficiently large to provide increased supplies commensurate with the assumed rise in drilling activities. An accelerated drilling effort during the 1980s could probably stabilize the present reserve level, but without such an effort both reserves and

[5] *BP Review of World Gas*, *BP Statistical Review of World Energy*, various issues.

production would certainly decline. In fact, the most likely assessment is that the depletion of existing wells will not be entirely compensated for by new developments. By the end of this decade gas availabilities in the USA may be lower than at the end of the 1970s.

The IEA production forecast for the USA is as follows (in tcf):[6]

1980	19.5 (actual)
1985	16.9 to 19.5
1990	16.1 to 21.1
2000	15.8 to 22.4

In this forecast much depends on reserve additions which are estimated as follows (excluding Alaska):

1980	—
1985	1.9 to 4.5
1990	5.2 to 10.0
2000	7.5 to 13.9

This means that the forecast assumes a very sharp drop in gas production from established reserves, in fact a decline from 19.3 tcf in 1980 to only 3.0 tcf in the year 2000. Such a significant dependence on expected reserve additions invites caution. Future US gas supplies may turn out to be smaller than the 'optimistic' IEA estimate for the 1990s, with the possibility of large gas imports during the 1990s if the rate of economic growth is high and if supply developments are disappointing. Increases in US gas imports will be met by Canada and Mexico in the first place, and should LNG imports from remote sources make a contribution it will largely depend on the price of oil in the 1990s.

Gas demand in *Western Europe* steadily declined between 1979 and 1982. This is similar to the demand trend in the USA mentioned in the preceding section, and many of the underlying causes are the same. It is certain however that the recovery that began in 1983–4 will be modest by past standards. Growth in future gas demand is likely, but accelerated expansion at high

[6] *Natural Gas: Prospects to 2000*, OECD/IEA, Paris 1982.

rates of increase is not. The movement in West European gas consumption was as follows (bcm):[7]

1973	144.3	1979	210.8
1974	165.4	1980	207.2
1975	172.4	1981	205.2
1976	183.8	1982	199.0
1977	190.9	1983	201.0
1978	196.2	1984	213.6

Gas supplies in Western Europe have three significant sources: the Netherlands, Norway and the UK. Although West Germany, Italy and France were important in the 1960s they were soon overtaken by the three new producers. In 1984 production was as follows (bcm):[8]

Netherlands	69.1
UK	35.4
Norway	29.6
West Germany	14.7
Italy	13.1
France	6.0

Gas trade is important in Western Europe. Natural gas imports by the main countries in 1984 were as follows (bcm):[9]

Austria	2.8
Belgium	9.0
France	23.5
Italy	20.1
Netherlands	2.8
Spain	2.0
Switzerland	0.4
UK	12.1
West Germany	35.5

Gas imports thus accounted for about half of total consumption, although some of these imports are intra-European:

— The *Netherlands* to Belgium, France, Italy, Switzerland and West Germany.
— *Norway* to Belgium, France, the Netherlands, the UK and West Germany.

[7] *BP Review of World Gas.* [8] *ibid.* [9] *ibid.*

Imports from outside Western Europe came from Algeria (18.1 bcm), the USSR (29.4 bcm) and Libya (1.1 bcm). In 1984 West European gas dependence on the outside world was therefore approximately one-quarter of total consumption. It would have been smaller had contracts signed with the USSR and Algeria in the 1970s not been based on an optimistic demand forecast which failed to materialise. Previously agreed increases in imports in the 1980s had to be honoured (at least in part), with the result that the decline in demand tended to penalize domestic production.

The European gas trade issue is very complex and has significant economic and political implications. The economic problem may be expressed as follows. On the one hand potential gas availabilities (both within and outside Europe) are considerable, and some observers think that in the long run they threaten to displace other fuels in the European energy market. On the other hand actual supply developments are sluggish because demand prospects are unattractive.

Since gas trade requires long-term contracts, importers and exporters are reluctant to commit themselves in the face of an uncertain future. Importers hope that gas might be supplied cheaply in the future, given availabilities and possible competition, while exporters do not want to make the necessary long-term and expensive investments unless they can secure a good price. Importers are inclined to wait in the expectation that exporters will be persuaded to lower their sights on prices; and exporters are reluctant to concede the lower prices lest they become tied to a bad deal for the next ten or fifteen years.

Ultimately, this apparent deadlock will have to be broken, since although Europe's indigenous natural gas reserves (in the Netherlands, Norway and the UK) now appear to be much more extensive than previously believed by most experts, nevertheless the need for additional imports from outside the region will increase. A recent Ruhrgas study estimates new annual imports at 35 bcm by the year 2000.[10] The forecast puts total West European demand in 1994 at 263 bcm, domestic production for domestic consumption at 113 bcm, intra-European trade at

[10] Bergmann, B., *Natural Gas in Western Europe – New Impetus in a Highly Competitive Market Environment*, paper presented to the Sixth Oxford Seminar, September 1984. See also Chapter 3 below.

58 bcm and imports from outside at about 90 bcm. These additional imports are likely to be provided by the present major suppliers, namely the USSR and Algeria. However, one can see an opportunity arising for new exporters such as Canada, Nigeria and perhaps the Gulf states.

It is important to emphasise that the volume of the gas import gap in the 1990s may well turn out to be different from this forecast, not only because the economic growth assumptions underlying the demand estimate are uncertain, but also because the export price policies of the main producers can alter the picture significantly. Should the producers opt to compete for a larger share of the market, they would succeed in displacing oil and further delaying the development of nuclear power. European dependence on gas imports would then increase. Similarly, we could expect that high gas prices would seriously constrain its future market share. However, it would be naive to believe that gas producers are poised for cut-throat competition. The investment costs of gas supply systems set a fairly high 'lower limit' to the price they can concede while still retaining a reasonable profit. The existence of large gas reserves does not in itself provide a sufficient incentive to sell the commodity cheaply; the problem of raising the finance for the infrastructure necessary to shift natural gas from the source to the market still has to be resolved. A strategy of 'dump today in order to create dependence tomorrow', a dependence that would give the producer some future market power and enable him to recoup in the future but survive current losses, is not viable given the long gestation period of such a policy. Neither the USSR nor Algeria, nor for that matter Norway, can afford to finance these expensive long-term strategies.

In fact there is much more pricing flexibility in oil than in gas, because at today's oil price the rent per barrel is much higher for oil than for gas. Rent, that is the gap between price and cost, is a key factor in inter-fuel competition, in as much as it provides the leeway for competition for market shares. Oil producers have the power, if they so desire, to disrupt the economics of all other fuels – coal, nuclear or gas. A long-term investor in any other energy source cannot feel secure (even with government backing) when the price/cost gap in oil is of the order of \$15–28. It is clear, therefore, that the future of gas in Europe depends on oil pricing policies *vis-à-vis* gas, and not on gas pricing policies. The room for manoeuvre on gas prices is limited; the room for manoeuvre on oil prices is very wide. In order to understand the future of the gas market, we need to anticipate oil prices rather than the possible gas pricing strategies of the USSR or Algeria. The gas producers require the support of a reasonable oil price; the oil producers, however, need not be concerned about competition from gas as long as they are able to use their pricing flexibility.

Gas trade in Europe also raises political issues. In this area a distinction must be made between the risk of interruption arising from disagreements about the implementation of contracts, and the risk of interruption arising from international hostilities. The former is a bargaining problem which is referred to as political, although in reality the disagreements may be about economic clauses such as prices and volumes of exports. The latter is a problem that may arise when the parties to a contract are in conflict over issues unrelated to the gas trade. An example of the first type of problem is the relationship between Nigeria and its customers; the second type of problem is usually mentioned when the Soviet Union is involved in a deal. It is more useful to look at the issue in terms of the difficulty of enforcing long-term contracts between sovereign states. A trading relationship can break down for a wide variety of reasons – technical, economic or political – and nothing is to be gained by explaining the failure in terms of the political standpoint of one party, with the other party, in all circumstances, an innocent victim.

By focusing attention on the contract rather than on the guilt of this or that party some constructive approaches may be developed. Not all contracts can be enforced by outside authorities, and they should therefore contain sufficient inducements for both parties to maintain the relationship in a world of changing circumstances, or sufficient flexibility to allow for mutual improvements whenever the situation permits. No ideal formula will ever be found. Cases will always arise where it pays one party to suspend the execution of a contract because benefits over time are perceived to be greater when the contract is suspended and then renegotiated than when it is implemented continuously according to the initial terms. Nevertheless, much effort and research needs to be devoted to improving our understanding of bargaining theory and our ability to devise contract terms that will lead to the establishment of mutually advantageous relationships between buyers and sellers of internationally traded gas.

Gas consumption in *Japan* is very small (37.2 bcm in 1984) compared, say, with the USA (514.6 bcm).[11] It is smaller than in West Germany, the UK or the Netherlands, and of the same size as in France. Domestic production in Japan is not significant and the gap is covered by imports from a variety of suppliers. In fact, Japan is more 'international' in its gas trade than any other importer. In 1984 its suppliers were Abu Dhabi, Alaska, Brunei, Indonesia (by far the largest) and Malaysia. Most other countries import from two or three sources only.

It appears that Japan is the market with the greatest potential for future LNG trade. The lack of significant domestic supplies, the growth of the economy, the environmental concern and the need to reduce concentration in its energy import dependence all militate in favour of growth in the LNG trade. Natural gas accounts for 9 per cent of primary energy consumption in Japan. Three-quarters of the imported gas is used for electricity generation; the rest is used for city gas and in industry. This pattern of use is very different from that in Europe and the USA where the domestic/commercial sector absorbs a large proportion of gas consumption.

The Pacific region will remain the privileged supplier of natural gas to Japan. Apart from the countries mentioned above, Australia and Canada will soon join the list of suppliers. Contracts with various consortia for loading at Grassy Point (Canada) and Withnell Bay (Australia) are pending finalisation. In addition, there is gas potential in Thailand, the South China Sea and Bangladesh, which may eventually represent new sources for the Japanese market. Moreover, a large expansion of Japan's gas demand would provide a welcome and significant inducement for the development of gas exports from the Gulf. However, there are other potential customers for Gulf exporters in South East Asia and the Pacific – for example, Korea, Taiwan and other industrializing countries.

[11] *BP Review of World Gas*, 1985.

Conclusion

The prospects for gas depend essentially on two factors:

(a) Economic growth.
(b) The future behaviour of oil prices.

Economic growth is the major determinant of increases in the demand for energy. Given growth, the world will want more oil and gas and perhaps more nuclear energy. Coal, however, is in a different class, being a fuel with a single market to conquer, that of steam-raising in the power sector. The driving force in coal demand is not so much growth but rather substitution; that is, relative prices and political concerns about import dependence on a single energy source.

Economic growth could eventually push the oil price up with the result that investments in high-cost gas supply systems would become more attractive. In other words, the higher the price of oil the greater the possibility of competition from gas. It should be emphasised that this presupposes the willingness of the gas producers to compete. Of course the possibility does exist today for a small lowering of gas prices, which would attract buyers and increase potential demand. Producers are naturally hesitant to try this at a time when the margins are relatively small, for fear of committing themselves for too long a period to an unfavourable price. However, the first difficulty can be solved by agreeing that contracts entered into freely at different dates are mutually independent. The second difficulty can be overcome by introducing a price escalation clause. Although this is common practice, it has caused problems in the past, usually because the escalation was automatic. A more flexible approach, which related the escalation to a number of parameters rather than time alone, or the price of oil alone, would be more useful.

The development and expansion of the international trade in gas are likely to be very slow for the rest of this decade but will gather momentum in the 1990s. The timing of investments in new supply systems may prove to be of crucial importance. While an early start would be wasteful of capital, a late start would have a high cost in delayed opportunities.

2 THE ENCROACHMENT OF NATURAL GAS IN THE OIL MARKET*

Aman R. Khan

Introduction

In the past five years, important changes have taken place in the energy sector world-wide as a result of the oil price shocks of the 1970s. The earlier fears of an impending shortage of oil supply have largely disappeared, replaced by a complacent expectation of adequate oil supplies until the end of this century. There are various reasons for the surplus of oil in the market today.

Consumers reacted to the drastic price increase of oil in 1979 by adopting energy conservation measures, thus cutting back the demand for oil. The high price of oil also provided an impetus to exploration and production, particularly in non-OPEC countries. This resulted in a substantial amount of oil entering the world market from Norway, the UK, Mexico, Egypt, Malaysia, etc., which cut into OPEC's share. In addition, exploration efforts were increased in the lesser developed oil-importing countries faced with high oil bills.

The glut of oil was further compounded by the world-wide industrial recession that followed the 1979 oil crisis. As the effect of the recession spread, the major consumers reduced their demand for oil and other energy supplies, with the reduction in petroleum demand being most pronounced. This effect was aggravated by the burgeoning strength of the dollar, which diluted the impact of the erosion of oil prices in the world market for those countries obliged to purchase their petroleum in dollars. In real terms, taking a basket of currencies of the major OECD industrial powers other than the USA, oil prices actually increased between mid-1983 and mid-1985

* Paper presented at the Oxford Energy Seminar, 1985.

by 9 per cent in the local currency equivalent, although in dollar terms the price has been essentially constant.

Another important reason for the reduction in oil demand was the replacement of oil by alternative energy sources, mainly coal and natural gas and to some extent nuclear power. The switch-over from oil to other fossil fuels was most noticeable in countries where such fuels were easily available at prices less than that of imported oil. Coal, which had taken a back seat in the energy spectrum, found an increasing role in the generation of electric power. However, the inherent disadvantage of coal, i.e. its contribution to environmental pollution, inhibits it from assuming a major share of the market vacated by high-priced oil.

On the other hand, natural gas, which is a clean-burning, non-polluting fuel does not suffer from this disadvantage. In countries where plentiful supplies of natural gas are available, the switch-over from oil to natural gas after 1979 was more pronounced, and natural gas took over sectors of the industrial, commercial and domestic markets once served by oil. With the present instability in the demand for oil and the uncertainty about its price, it is pertinent to examine the outlook for natural gas as a displacement fuel and the share of the traditional oil markets that gas can assume.

Background to Oil Displacement

At the time of the first oil crisis in 1973, the consumption of crude oil in the non-Communist world amounted to 46 mb/d. The relation between OPEC's marker crude oil price and consumption in the non-Communist world is shown in Figure 2.1 and indicates one of the main reasons for the reduction in demand for OPEC oil. With the price increase in 1973, this consumption declined in the next two years, and as oil prices remained relatively stable from 1974 to 1978, oil consumption increased to 51 mb/d. As oil prices soared from $13.48 per barrel in early 1979 to a record high of $34 per barrel by the end of 1981, oil consumption in the non-Communist world dropped drastically to 45 mb/d in 1982 and remained at that level until 1984.

With the higher price of oil, consumers were forced to seek

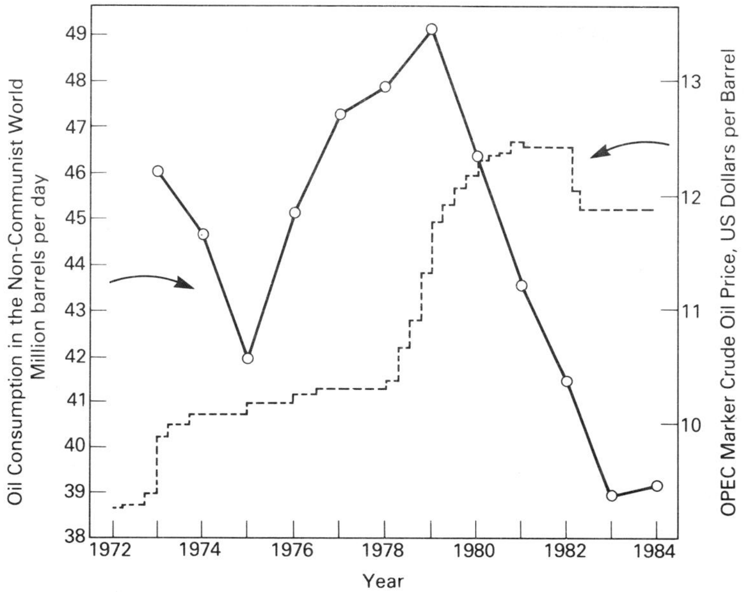

Figure 2.1 Relation of Non-Communist World Crude Oil Consumption to OPEC Marker Crude Oil Price.

remedies by curtailing energy demand and substituting for oil. There was also an increased effort to develop non-OPEC sources of oil. In the decade following 1973, the number of oil-producing countries in the world increased by a third and, of these, the UK, Norway, Egypt and Mexico ranked as major exporters, cutting into OPEC's traditional markets. OPEC's share of world oil exports has declined from 80 per cent in 1979 to less than 60 per cent in 1984, and it is anticipated that it will decline further in 1985.

The trend towards conservation of energy in the major oil-consuming countries also had a decisive effect on the demand for oil and, with oil viewed as an expensive and unstable energy source, the thrust to substitution for oil was inevitable. Coal and natural gas, the other two most commonly used fossil fuels, showed substantial growth in their share of the energy spectrum. The world consumption of coal increased from 2,210 million tonnes in 1973 to 2,765 in 1982, an increase of about

7 mboe/d. Natural gas consumption world-wide increased from 43 tcf in 1973 to 57.6 tcf in 1984, or a difference of over 7 mboe/d. This is quite significant in terms of total world oil consumption and emphasises the fact that both coal and gas have made a dent in the demand for oil and that while the demand for oil is depressed, the demand for gas shows a healthy growth.

In the case of natural gas, world proven reserves at the end of 1984 amounted to 3,400 tcf, an increase of about 6.5 per cent over the previous year. This is approximately 596 billion boe or almost 90 per cent of actual oil reserves. The world reserve life index for natural gas is fifty-six years as compared with thirty-four years for crude oil, and the prospects for developing additional reserves of gas are very bright. Most experts agree that a rise in world gas consumption is inevitable and that this will occur largely at the expense of oil. The oil market sectors most susceptible to displacement are the industrial and commercial loads. The transportation sector is unique in that reliance on liquid fuels will continue, and the use of natural gas as a vehicular fuel, either directly (CNG) or as alcohol or synthetic gasoline, is still at an early stage and has not gained wide acceptance. Two factors have an important bearing on the displacement of oil by natural gas – availability of long-term supply and competitive price – and it is in this context that one should examine the present and potential inroads that gas has made and will make into the oil market. As stated earlier, the supply of gas is more assured on a long-term basis than that of oil. However, it should be recognised that this is on a regional basis since the world trade in gas does not have the same flexibility as the international oil trade. Insofar as inter-fuel competition is concerned, natural gas is priced lower than oil products to the consumer in most world markets with a few exceptions, where national policies regulate all energy prices.

The Industrial, Commercial and Domestic Markets

Discounting the transportation sector where oil has maintained its traditional role, coal and gas have posed a serious threat of breaking into the industrial sector, and natural gas has made significant advances into the commercial and domestic

Figure 2.2 Residual Oil Consumption in the Non-Communist World.

markets. Consumption of residual and distillate oils which meet market demands in these sectors has declined significantly since 1979. The major consumers for these oil products in the non-Communist world are the USA, Japan and Western Europe, and Figure 2.2 shows the pattern of *residual oil* consumption in the non-Communist areas over the period 1973–82. The principal loads for residual oil are power generation and industry and it can be seen that consumption declined from 13.7 mb/d in 1973 to 12.1 mb/d in 1975. As crude oil prices remained relatively stable for the next three years, residual oil demand increased gradually to its previous level. With the second crude oil price increase in 1979, residual oil demand dropped drastically to less than 10 mb/d by the end of 1982. This segment of demand for power generation and industry was taken over by coal and natural gas, although some of the decrease is attributable to successful measures in energy conservation.

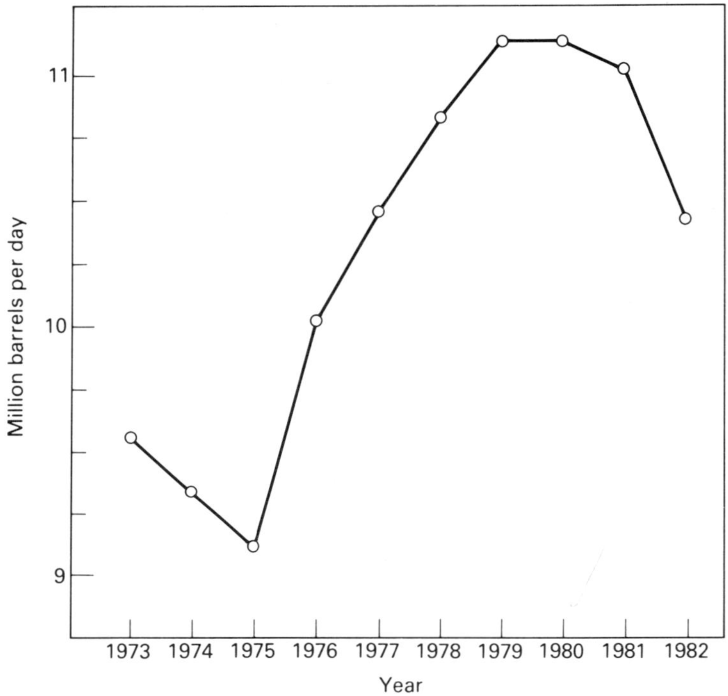

Figure 2.3 Distillate Oil Consumption in the Non-Communist World.

In the case of *distillate oil* consumption, the decline was less marked, as shown in Figure 2.3. The market for this oil product is mostly commercial and domestic, and fuel-switching at the individual installation is probably less economically attractive. Again, from 1973 to 1975, distillate oil consumption in the non-Communist markets decreased from 9.6 to 9.1 mb/d. Subsequently, consumption increased to 11.1 mb/d by 1979 and thereafter declined to 10.4 mb/d by 1982. The differential in consumption was largely picked up by natural gas.

The major reason for fuel-switching from oil to gas was the relatively lower price of gas to the consumer. In the case of the industrial market both residual fuel oil and light fuel oil are used and natural gas had a substantial price advantage after 1979. In the domestic and commercial market, light fuel oil is used and the price advantage for gas is somewhat less because the distribution and marketing costs are higher for gas than for oil.

Table 2.1 presents a comparison of average energy costs to the consumer over the period 1978–83 for selected OECD countries (UK, USA, West Germany, France, Italy and Japan). These countries are major consumers of oil and gas in the industrial and domestic markets. In all cases except Japan, the price of natural gas to the industrial consumer is well below that of light and heavy fuel oil. For the domestic market the pattern is the same, with natural gas prices less than those of light fuel oil. However, it should be noted that as crude oil prices declined in 1983 to $29 per barrel, the price of fuel oil to the domestic consumer dropped and was competitive with natural gas. An exception to this price differential between oil products and gas is Japan. Practically all of Japan's natural gas is imported in the liquefied form (LNG) from Alaska, Indonesia, Brunei, Malaysia and Abu Dhabi. The price of LNG delivered to terminals in Japan (c.i.f.) is tied in with a particular crude oil or a basket of crude oils and the pricing formula is indexed to inflation and currency exchange rates. To the delivered cost must be added the costs of regasification and distribution, which results in a higher price than that of oil products delivered to consumers from local refineries. Nevertheless, in keeping with the country's policies of reducing pollution and providing a flexibility and stability of energy supply, natural gas has a share of the market and plans call for increasing gas imports.

The decline in oil's share of the industrial and domestic market demonstrates the fact that as long as natural gas is competitive at the point of usage, it will continue to make inroads and increase its share of energy consumption. Although it is difficult to identify exactly how much of the industrial, commercial and domestic market was lost to gas world-wide, an examination reveals that in most regions of the world in both the oil-exporting and importing countries, there is an increasing trend towards gas utilisation and replacement of oil where it is practically and economically feasible.

Table 2.1: Energy Costs to Consumer Including Taxes. Dollars per Barrel of Oil Equivalent.

	Domestic		*Industrial*		
	Fuel Oil	*Natural Gas*	*Light Fuel Oil*	*Heavy Fuel Oil*	*Natural Gas*
UK					
1978	25.83	19.55	20.29	15.36	12.14
1979	35.91	22.52	30.45	21.29	14.82
1980	53.97	29.92	47.59	32.20	22.12
1981	55.15	31.58	47.17	33.89	23.48
1982	55.86	33.57	44.65	31.28	21.83
1983	50.40	32.43	39.90	29.53	19.61
USA					
1978	20.58	14.65	16.51	11.78	8.04
1979	28.48	18.18	24.15	15.98	13.17
1980	40.45	22.29	35.07	21.05	14.36
1981	50.19	24.05	45.36	28.51	17.56
1982	49.48	29.07	44.31	25.86	21.15
1983	45.15	34.43	37.38	25.74	24.45
West Germany					
1978	24.49	34.71	21.84	15.59	14.54
1979	43.81	38.25	38.64	20.62	16.42
1980	54.52	45.20	48.09	28.01	21.32
1981	50.32	43.72	44.52	32.19	22.23
1982	51.66	49.36	45.78	28.48	27.47
1983	44.94	47.08	39.48	26.89	28.39
France					
1978	30.03	37.39	25.54	13.99	13.40
1979	39.19	43.43	33.31	18.33	15.56
1980	58.30	56.15	49.69	27.39	23.88
1981	58.80	55.12	49.98	30.07	25.42
1982	57.67	53.35	48.72	20.57	25.82
1983	54.73	59.25	46.20	27.49	24.74
Italy					
1978	27.93	25.25	24.70	13.30	12.20
1979	35.62	31.01	31.21	17.85	17.10
1980	57.67	37.85	50.61	27.41	25.54
1981	51.37	38.19	44.65	29.76	26.85
1982	56.49	40.58	49.27	26.55	28.96
1983	57.12	40.30	49.69	25.65	24.97

Table 2.1 (*cont.*)

	Domestic		*Industrial*		
	Fuel Oil	*Natural Gas*	*Light Fuel Oil*	*Heavy Fuel Oil*	*Natural Gas*
Japan					
1978	32.34	66.58	n.a.	n.a.	52.95
1979	37.80	64.98	n.a.	n.a.	51.13
1980	62.37	78.60	51.79	43.37	57.97
1981	68.04	91.88	58.93	39.99	66.40
1982	66.99	83.79	67.41	38.31	60.42
1983	66.91	86.92	71.40	41.40	61.67

Source: Energy Information Administration, US Department of Energy

Industrial Fuel-switching

Data are not readily available on the inroads that natural gas has made in specific marketing sectors in the period following the two oil price increases. A parallel to this displacement of oil may be seen in the development of the coal industry in the decade after World War II. At that time, the major energy source was coal. However, in the USA, the continued agitation by the labour unions for increased wages and persistent threats of strikes resulted in higher prices for coal and the consumer response was a switch to oil, which was readily available, easily transportable and relatively cheap. The market lost by coal to oil was never fully recovered and coal was displaced after a century as a prime energy source. Inter-fuel competition resulted in both oil and gas displacing coal in order to meet industrial and domestic heating requirements. It now appears that a repetition of this situation has taken place, with oil being viewed as a costly and unreliable fuel. Needless to say, the consumer response is to switch to cheaper alternative fuels.

Fuel-switching has been more pronounced in the industrial sector where the market is most vulnerable to inter-fuel competition. For fuel-switching to take place, there are certain elements which are prerequisite. In countries which are dependent on oil imports, it may be a national energy policy which dictates a diversification of energy sources and a reduction of oil imports. Secondly, it is desirable that the alternative energy source is indigenous. There must be assured long-term supply and the price should be fairly stable and not subject to wide variations as in the case of oil.

In the case of gas as an alternative to residual oil in the industrial market, there are certain factors to be taken into consideration. The most important is the gas/residual oil price ratio. For each application, there are also specific technical circumstances to be considered as well, such as the industry's end-use requirements, storage and handling facilities and environmental regulations. Since the second oil price increase, the industrial market in the USA has been characterized by price competition with fuel-switching between residual oil and gas. A comparison of the volume of fuel-switching between these two fuels for 1983 and 1984 indicates that where the gas/residual oil price ratio was below or near parity, the fuel-switching was in favour of gas. With the shut-down of refineries and the move towards hydrocracking of residual oil to maximize production of distillates, the price of residual oil has increased, thereby giving gas a further advantage.

A recent analysis by the American Gas Association assesses the prospects of increased use of gas to displace oil in today's market. There is sufficient short-term supply and flexibility in delivery systems to displace 340,000 b/d of oil in stationary uses within one month if an oil supply disruption occurred. Within one year of such a disruption, about 730,000 b/d of oil could be displaced by gas. In other words, the potential for gas to displace oil in these markets could allow a significant cut-back in oil imports. However, there are certain constraints on natural gas usage in industry in the USA, the major one being the Fuel Use Act which prohibits the use of gas in new electric power plants and large industries. Thus, for the time being, the gas industry in the USA can only focus on fuel-switching in existing industrial loads. Nevertheless, about 60,000 b/d of oil was lost to gas in 1983 in the electric utility sector and about 67,000 b/d of oil was displaced in 1984 in both the industrial and electric utility sectors.

Natural Gas Demand

At the recent sixteenth World Gas Conference held in Munich in 1985, a Task Force Group on World Gas Supply and Demand presented estimates of potential gas demand in the world assuming a 'moderate' oil price scenario. This assumes that oil prices will increase to $30 per barrel by 1990, $40 by the year 2000 and $55 by 2020, reflecting a stable energy market until 1990 and a modest growth in energy demand thereafter. The basic elements in forecasting the potential gas demand are large gas reserves, a growing energy consumption, a need in many countries to reduce oil imports and diversify energy sources and the recognition that gas is a non-polluting and efficient fuel.

Table 2.2: World Gas Demand, by Region. Million Barrels of Oil Equivalent per Day.

	Actual			*Potential Economic*			
	1980	*1982*	*1983*	*1990*	*2000*	*2010*	*2020*
Africa	0.14	0.27	0.32	0.68	1.28	1.87	2.73
Asia	0.59	0.59	0.68	1.46	2.10	2.46	2.60
Eastern Europe	7.52	8.61	8.88	11.25	15.58	16.86	18.22
JANZ*	0.64	0.73	0.77	1.46	1.73	1.78	2.00
Latin America	1.09	1.18	1.09	1.37	1.91	2.37	2.87
Middle East	0.41	0.50	0.55	1.32	2.19	2.46	2.73
North America	10.52	9.84	9.34	10.98	12.07	13.03	14.22
Western Europe	3.64	3.37	3.51	4.51	4.51	4.69	4.83
World Total:	24.55	25.09	25.14	32.67	41.37	45.52	50.20

* Japan, Australia, New Zealand

Source: Report of Task Force, World Gas Supply and Demand, 1983– 2020. 16th World Gas Conference, Munich 1985.

Table 2.2 presents the gas demand in various regions of the world over the period 1980–2020 in million barrels of oil equivalent per day. The major consuming areas are North America, Western Europe and Eastern Europe. Until 1990, the projected average yearly growth in demand is 3.8 per cent followed by a 2.4 per cent annual growth rate until the end of the century. Of course, this assumes that oil prices will be $40 per barrel by that time. With a rapid increase in oil prices, the growth in demand for gas will be higher than that shown in Table 2.2 for the year 2000. A breakdown of world gas demand by market sector is presented in Table 2.3. For the industrial sector, there is a projected annual growth rate of 3.9 per cent over the period 1983–90 and this declines to 3 per cent over the next decade. With respect to the use of natural gas in the generation of electric power, the annual growth rate in demand is 3.7 per cent for the period 1983–90, but this declines to about 1 per cent per annum in the following ten years as nuclear power builds up. The residential and commercial loads show an average growth rate of 3.2 per cent per annum from 1983 to 1990 and a modest rate of 2 per cent per annum for the period 1990–2000.

Table 2.3: Gas Demand, by Market Sector.
Million Barrels of Oil Equivalent per Day.

	Actual			*Potential Economic*	
	1980	*1982*	*1983*	*1990*	*2000*
Residential and Commercial	6.78	7.06	7.02	8.79	10.61
Industrial	10.52	10.11	10.16	13.30	17.95
Electric Power Generation	4.46	4.74	4.69	6.06	6.70
Chemical and Other Usage	2.79	3.18	3.27	4.52	6.11
Total	24.55	25.09	25.14	32.67	41.37

Source: Report of Task Force, World Gas Supply and Demand, 1983–2020. 16th World Gas Conference, Munich 1985.

By contrast, forecasts by the oil industry for oil consumption are far less optimistic and show lower growth rates with a decline in oil's share of energy supply by the end of the century. For *Western Europe* the share of oil in the total energy supply declined from 62 per cent in 1973 to 48 per cent in 1984, and is forecast to drop to 42 per cent by 1990 and 38 per cent in the year 2000. Although the EEC forecasts a faster growth for coal and nuclear power, it is expected that natural gas will increase its share of the energy market to 24 per cent by the year 2000.

In France, oil demand for electric power generation has been substantially reduced and will continue to decline as more nuclear and coal-fired plants come on stream. Natural gas consumption is forecast to grow faster than total energy consumption as gas replaces oil in the industrial and space heating loads.

In Italy, the share of oil in primary energy demand decreased from 75.6 per cent in 1973 to 59.2 per cent in 1984 and is forecast to decrease to 53.7 per cent in 1990. On the other hand, natural gas increased its share from 14.3 per cent in 1973 to 18.7 per cent in 1984 and is expected to rise to 20.1 per cent by 1990. In 1984, about 60 per cent of the natural gas consumed in Italy was imported and total consumption was over 1 tcf. It was a major source of industrial and residential energy and it is anticipated that total consumption will be 1.6 tcf or 770,000 boe/d by the year 2000. The country's present energy policy seeks to diversify energy supplies and particularly reduce dependence on imported oil by increasing the shares of coal, nuclear power and natural gas and by stepping up the development of national energy sources.

The West German energy programme calls for increased conservation, more reliance on nuclear power, coal and natural gas and a declining use of oil. West German oil consumption declined from 3 mb/d in 1973 to 2.3 mb/d in 1984 and a very slight increase to 2.4 mb/d by 2000 is expected. Natural gas provides over 15 per cent of primary energy demand. Residential and commercial demand account for 38 per cent of total gas sales with another 31 per cent for gas usage in industry. The use of natural gas in power stations is projected to decline, but increases in natural gas consumption in the industrial and residential sectors are forecast with an overall growth in consumption of 1.2 per cent per annum to the year 2000.

The largest gas consumption in Western Europe is in the UK where natural gas meets about 23 per cent of primary energy demand. About 65 per cent of total gas consumption is for the residential and commercial market with another 30 per cent used in industry. Natural gas is not used for power generation, a market which is served by nuclear, coal and fuel oil. The 1984 coal strike reduced coal consumption and resulted in an increase of fuel oil consumption by 260,000 b/d, but this has been reversed this year and oil consumption is forecast to be 1.5 mb/d by 1990 with a growth of 1 per cent per annum for the next ten years. Conversely, natural gas consumption is expected to increase steadily with a growth rate of 1.6 per cent per annum to the year 2000.

The largest natural gas-consuming region in the world is *North America* where gas accounts for 30 per cent of primary energy. In the USA, gas provides 42 per cent of the energy requirements in the residential and commercial market, about 41 per cent of industrial energy and 12 per cent as fuel for power generation. With the increased emphasis on energy conservation, it is expected that there will be some decline in the share of residential and commercial gas consumption for North America, although in absolute amounts gas sales will increase by about 14 per cent over present sales by 1990. Gas usage in power generation is expected to decline, whereas for the industrial sector a growth of 2.7 per cent per annum is forecast till 1990 followed by about 1.3 per cent per annum till 2000. Thus, the biggest potential for gas is the industrial market including on-site co-generation, although it faces competition from fuel oil with declining oil prices. The Gas Demand Committee of the American Gas Association has prepared forecasts on gas demand in various sectors for varying assumptions of oil prices and energy demand. For the base case (with moderate oil prices and moderate overall energy demand growth) the outlook for gas demand by sectors is presented in Table 2.4. Overall consumption is expected to increase by only 1 per cent per annum with the major growth being in industrial markets at the expense of oil.

In the *Middle East*, there is an increasing awareness of the potential for natural gas as a feedstock for new industries and as a fuel for power plants and desalination units. Gas utilisation in the Gulf countries increased from 1.84 bcf/d in 1972 to almost 4 bcf/d by 1982. Practically all of this was associated gas and in 1983, as oil production cuts were made, serious shortages of gas

Table 2.4: Outlook for Gas Demand by Sectors in the USA. Trillion Cubic Feet.

	Actual			*Forecast*	
	1981	*1982*	*1983*	*1990*	*2000*
Residential	4.9	5.1	4.8	5.0	5.1
Commercial	2.3	2.4	2.3	3.0	3.9
Industrial	8.3	6.9	6.7	7.9	9.0
Electric Power Generation	3.8	3.3	3.0	3.2	3.0
Pipeline Fuel and CNG	0.6	0.6	0.6	0.8	1.1
Total	19.9	18.3	17.4	19.9	22.1

occurred with curtailments in electric power supply and in the production of liquefied petroleum gases. Consequently, steps are being taken in Saudi Arabia, Abu Dhabi and Qatar to develop supplies of non-associated gas. The largest non-associated gas field is located offshore Qatar and plans call for production of 2.4 bcf/d. The projected demand for gas by 1990 in the region is estimated to be two-and-a-half times present consumption and this will be doubled in the next decade. Unlike the OECD countries, about half of this demand will be for power generation with the remainder for desalination plants and as a process feedstock.

In *Latin America*, the major gas-consuming countries are Mexico, Venezuela and Argentina. In Mexico, the largest single consuming sector is the industrial market, and gas is forecast to meet half the needs of this sector. Although the residential demand is small, this is expected to increase and the national policy is to reduce internal oil consumption by gas displacement, thereby providing more oil for export and as a foreign currency earner.

A similar policy is being followed in Venezuela where most of the gas produced is associated gas. With the decline in oil production, plans are under way to develop non-associated gas reserves in the eastern region. The Nurgas project is designed to supply 1 bcf/d from the Oriente to the western part of

Venezuela and replace oil usage in the industrial, commercial and domestic loads. The first phase of the pipeline is from Anaco to Morón and subsequently in the second phase this will be expanded to Rio Seco thereby assuring continuous supply of gas which is not subject to the vagaries of oil production.

The Government of Argentina is encouraging a major switch to natural gas as a feedstock for petrochemicals and for power generation. From 1974 to 1984, gas sales to the industrial market increased from 140 to 212 bcf per annum. The share of gas in domestic energy demand increased from 22.8 per cent to 31 per cent during the same period. There is also potential for increased gas consumption in Bolivia, Brazil, Colombia, and Chile, and for Latin America as a whole, gas demand in the industrial sector and for power generation is expected to double by the end of the century.

In *Africa*, natural gas is forecast to be the fastest growing energy source. Presently, oil accounts for 49 per cent and natural gas for 9 per cent of total energy demand. By the year 2000, gas will increase its share to 18 per cent and oil will decline to 40 per cent. The major consumption of gas will be in the North African countries (Algeria, Egypt, Tunisia and Libya) and Nigeria. However, there has been considerable activity in other parts of the continent and the potential for gas development is very promising in the Ivory Coast, Sudan, Angola, Tanzania and the Cameroons.

In *Asia* (excluding Japan), the demand for gas in 1983 was estimated as 1.4 tcf per annum and accounted for 9 per cent of energy demand. A healthy annual growth rate of 9.2 per cent is forecast with a market share of 16 per cent predicted by 2000 amounting to over 4 tcf annually. Again, the share of oil in the energy demand in the region is forecast to decline from 49 per cent to 32 per cent over the same time period although present consumption of 3.4 mb/d will increase on average by 2 per cent per annum. About 75 per cent of the gas demand in the region is consumed in China, Pakistan and Indonesia. Other countries with developing gas demand are Bangladesh, Malaysia, Taiwan, Thailand and India. In Pakistan, the demand for gas has increased beyond the capabilities of the present transmission network, and although additional supplies are being developed, plans call for new power stations and industries to use coal in order to conserve gas for those markets where it can reduce oil imports. The replacement of oil by natural gas is very noticeable in Thailand. Fuel oil consumption for power generation has decreased from a share of 60 per cent in 1981 to 30 per cent in 1984 and is projected to decline to 3 per cent by 1990. Natural gas has increased its share from 10 per cent in 1981 to 29 per cent in 1984 and the target for 1990 is 60 per cent. In Malaysia, present usage of gas is limited to LNG exported to Japan. However, with the construction of the offshore gas gathering system and trans-peninsular pipeline, natural gas is forecast to provide almost 40 per cent of the energy demand by 1995. Similarly, in Indonesia, apart from the substantial LNG exports, gas is estimated to assume an increasing share in meeting primary energy requirements.

In the *Pacific Basin*, the major consumers of gas are Australia, New Zealand and Japan with South Korea as a future importer of LNG. As stated earlier, Japan is a major importer of gas and the prognosis is that gas will increase its share of total energy consumption from 9 per cent in 1984 to over 11 per cent by 2000 with an average growth rate of 3.5 per cent per annum. By contrast, during the same period, oil consumption will have an average annual growth rate of only 0.3 per cent and will account for 47 per cent of total energy consumption in 2000 as compared with 60 per cent in 1984. However, the displacement of oil will be shared by nuclear, coal and gas.

Both Australia and New Zealand have plans to maximize the use of local non-oil energy reserves and decrease oil imports. Thus, oil's share of total energy consumption for the two countries is expected to decline from 42.5 per cent in 1984 to 33.5 per cent by 2000. However, public sentiment precludes nuclear power as a replacement for oil and increased use of coal and natural gas is planned. In Australia, 43 per cent of gas demand is for industrial usage and 23 per cent for electricity generation, and gas is very competitive relative to oil. The proportion of primary energy supplied by gas is expected to increase from its present level of 14 per cent to 19.8 per cent by 2000.

In New Zealand, which has abundant natural gas resources but very limited oil resources, the Government is committed to a policy of replacing oil with indigenous gas in all sectors. In fact, the country is probably the only one where the transportation sector is designated for substitution of motor gasoline by CNG or synthetic gasoline derived from gas via the methanol route. The latter substitutes will account for as much as 40 per cent of automotive fuels by 2000.

Based on the above review, the potential for gas demand in various regions of the world appears very promising. Gas consumption is expected to continue increasing world-wide and a substantial portion of this increase will be at the expense of oil if petroleum prices remain at the high levels attained in the period 1980–85.

Conclusions

Natural gas has made significant gains in displacing oil in some of its traditional markets and this trend may continue in the next decade depending on the future behaviour of oil prices and the rate of growth of the world economy. Natural gas has certain advantages over oil in markets where it competes directly. There are ample supplies of gas in the world and its reserve life index is much higher than that of oil, with prospects for additional reserves world-wide. It is a non-polluting fuel unlike fuel oil or coal, which contain sulphur, and it is usually lower priced than oil products at the burner-tip. However, in addition to its natural physical advantages, it is the lack of a fixed world price for natural gas, in contrast to oil, which has allowed gas the flexibility to encroach on oil in particular market segments. For, as in the case of coal after World War II, it is the uncertainty of oil's availability at a reasonably stable price that has prompted the consumers to seek a substitute fuel. Until stability of price and supply is restored in the oil industry, the potential for encroachment on oil by gas will continue in the markets where gas offers a least-cost option, i.e. in the industrial and domestic markets.

It is clear that natural gas can and will displace oil if market conditions are right. The long-term availability of gas is much better than that of oil, and availability presages price stability.

After many years of artificially structured prices, natural gas has shown that it can follow energy prices set by a free market, at least in those regions where it can be made available via pipeline. In many developing countries faced with the burden of higher oil import bills, it is the only indigenous fuel available in sufficient quantities to meet the growing demands of an industrial economy. These prospects are largely dependent on the price advantage relative to the fuels it will displace. Therefore, for gas to maintain its competitive advantage and to continue to develop its deserved role in the market, it is essential to remove the restraints and regulations that restrict its growth. Its price will have to be responsive to the needs of the consumer.

3 NATURAL GAS IN WESTERN EUROPE: MODEST GROWTH IN A BUYERS' MARKET?*

Burckhard Bergmann

Introduction

In late June 1985, the International Gas Union held its sixteenth World Gas Conference. The Conference, which was held in Munich, not only attracted a greater number of delegates than previous congresses of the IGU, but also imparted an atmosphere of reassurance and confidence, demonstrating a strong belief in the future of the world's gas markets.

This paper briefly reviews energy market trends since 1979–80, analyses today's gas scene and, finally, outlines the long-term perspectives of supply and demand and discusses whether they justify or mitigate the optimistic mood of Munich.

Energy Market Transformation between 1979 and 1982–3

Whereas the growing share of gas in the primary energy market was not adversely affected by the first oil crisis – boosted energy prices even enhanced the viability of new import projects – gas was unable to escape the repercussions of the oil price rises in 1979–80 (see Figure 3.1). The exposure of gas to the negative effects of the depression in the market was due to various factors. Two reasons were of particular significance:

(a) In absolute terms, the increase in the price of oil was more than twice as high as in 1973–4. This triggered a major economic recession and led to a slump in energy demand until 1982. Total energy consumption fell by 8 per cent over this period, whereas the reduction had been only 5.5 per cent between 1973 and 1975.

* Paper presented at the Oxford Energy Seminar, 1985.

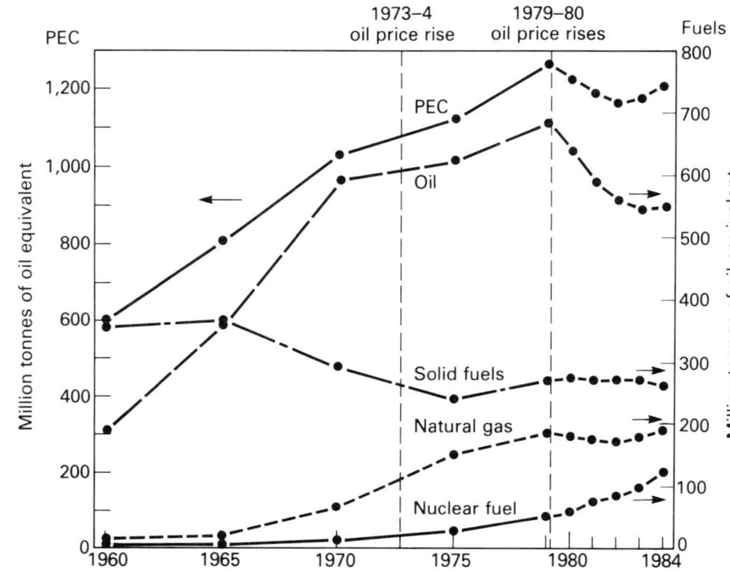

Figure 3.1 The Transformation of the West European Energy Market. 1960–84.

(b) The price of gas, pegged to the price of oil products, rose in the wake of the oil price rises and, for the first time, gas was unable to compete with coal in price terms. It also found it difficult to hold its own against low-cost electricity generated by nuclear and hydroelectric power stations or coal-fired power plants.

Figure 3.2 compares economic data and consumption data for different sources of energy in Western Europe from 1979 to 1984. Whereas the gas industry slightly increased its overall energy market share, it still hardly surpassed the oil industry until 1982–3, when the downturn in heavy fuel oil demand took on almost dramatic proportions, dropping by some 39 per cent between 1979 and 1983. Gas oil sales fell by approximately 26 per cent in the same period. Electricity was the only sector of the energy industry that succeeded in increasing its sales despite the marked decline in energy consumption, but the coal industry was able to maintain hard coal sales at earlier

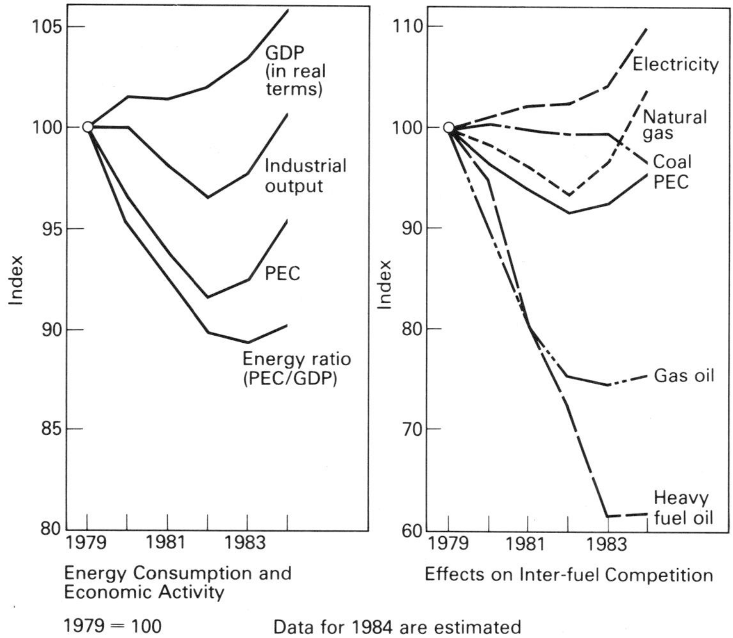

Figure 3.2 Changes in Western Europe since 1979.

levels despite the slack demand from the iron and steel industry and, as a result, it expanded its share in primary energy supplies.

Various factors explain why the transformation of the energy market did not have the same impact on all sources of energy:

(a) The prices of the various fuels changed at different rates, mainly because the options available to the different sectors of the energy industry in responding to the changing market environment were different owing to different cost developments and fixed-to-variable cost ratios.
(b) Demand developments were not uniform across the various market sectors, in which the individual fuels play different roles.
(c) Environmental protection requirements affected the various sources of energy differently.
(d) Confidence in the security of oil supplies changed.

Table 3.1: Final Energy Consumption in Western Europe, by Market Sector. 1979–84. Percentage Shares.

	1979	*1980*	*1981*	*1982*	*1983*	*1984*
Industrial	36	35	34	34	33	34
Transport	23	24	25	26	26	26
Residential and Commercial	41	41	41	40	41	40
Total (mtoe)	850	815	790	780	775	790

Note: 1984 figures are tentative.

Source: Ruhrgas

Final energy consumption by market sectors in Western Europe is shown in Table 3.1. *In toto*, final energy use declined by 9 per cent between 1979 and 1983. This decrease was mainly caused by a significant fall in final energy demand by industry and by residential and commercial users, although slightly more automotive fuel was used.

The shares held by the different sources of energy in the sectors of the energy markets in major West European nations for 1983 are summarized in Table 3.2. Although the patterns are by no means consistent, the Netherlands and, to some extent, the UK are atypical cases because of the high domestic gas output. In West Germany, Italy and France, on the other hand, gas holds similar shares of 20–28 per cent in the residential and commercial energy market and 21–25 per cent in the industrial market. Figure 3.3 compares price and consumption trends for West Germany. In spite of stable energy demand in the transport sector, oil was worse hit than other sources of energy, since oil products prices rose more steeply and confidence in the reliability of oil supplies was shaken. Although the development of gas prices has, on the whole, been in keeping with those of gas oil and heavy fuel oil, gas was able to escape the fate suffered by oil. The diversification of sources of energy led to increased confidence in the security of supply, and the environmental compatibility of the fuel became increasingly important.

The contraction of heavy fuel oil demand in the industrial

Table 3.2: Percentage Shares of Fuels in Energy Consumption of West European Countries. 1983.

	France	*Italy*	*Netherlands*	*UK*	*West Germany*
(*a*) *Residential and Commercial*					
Coal	6	1	1	15	6
Oil	49	52	11	16	43
Gas	22	28	74	46	20
Electricity	23	18	14	23	19
District Heating	0	0	0	0	4
Other	0	1	0	0	8
Total (mtoe)	50	34	18	55	73
(*b*) *Industrial*					
Coal	18	13	9	17	25
Oil	34	34	13	26	19
Gas	21	25	55	35	23
Electricity	22	24	20	18	24
District Heating	0	0	1	0	1
Other	5	4	2	4	8
Total (mtoe)	34	31	12	37	53
(*c*) *Power Stations*					
Coal	23	15	24	70	63
Oil	6	62	9	9	3
Gas	2	9	58	1	8
Nuclear Power	67	5	7	20	18
Other	2	9	2	0	8
Total (mtoe)	56	32	13	66	84

Source: Ruhrgas

and power station sectors can be attributed in the main to the losing battle fought by heavy fuel oil against highly competitive hard coal, making it an increasingly attractive proposition to transform residual fuel into lighter products.

Table 3.3 shows the development of natural gas sales to the different market sectors in Western Europe, while Figure 3.4 shows rates of market penetration. As the graphs show, the gas industry was able to expand its market share in the residential

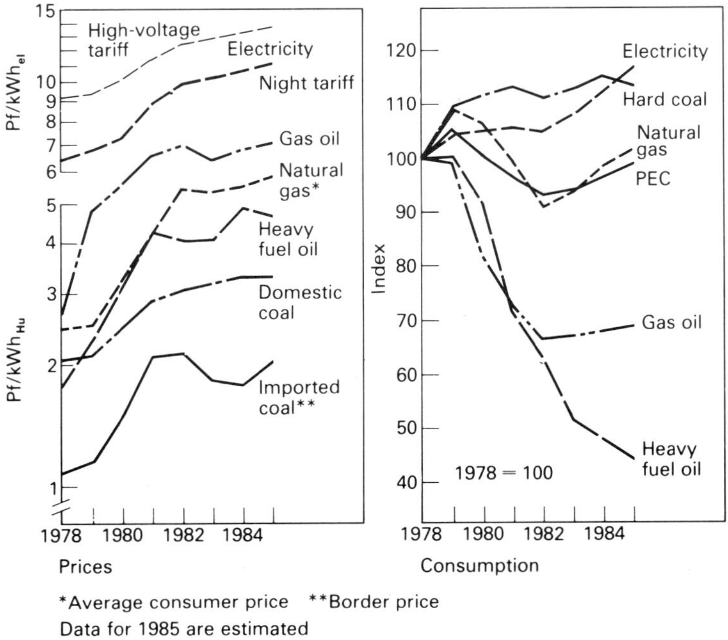

Figure 3.3 Price and Consumption Indices for Various Sources of Energy in West Germany since 1978.

Table 3.3: Natural Gas Consumption in Western Europe, by Market Sector. 1979–84. Percentage Shares.

	1979	*1980*	*1981*	*1982*	*1983*	*1984*
Residential and Commercial	40	42	44	45	45	45
Industrial	33	33	32	32	30	30
Power Stations	16	14	13	13	13	13
Other	11	11	11	10	12	12
Total (mtoe)	184	181	177	171	178	190

Note: 1984 figures are estimates.
Source: Ruhrgas

and commercial sector. It also held its own in the industrial sector, in spite of a decrease in volume sales, although a major set-back was suffered in the power station market, where sales

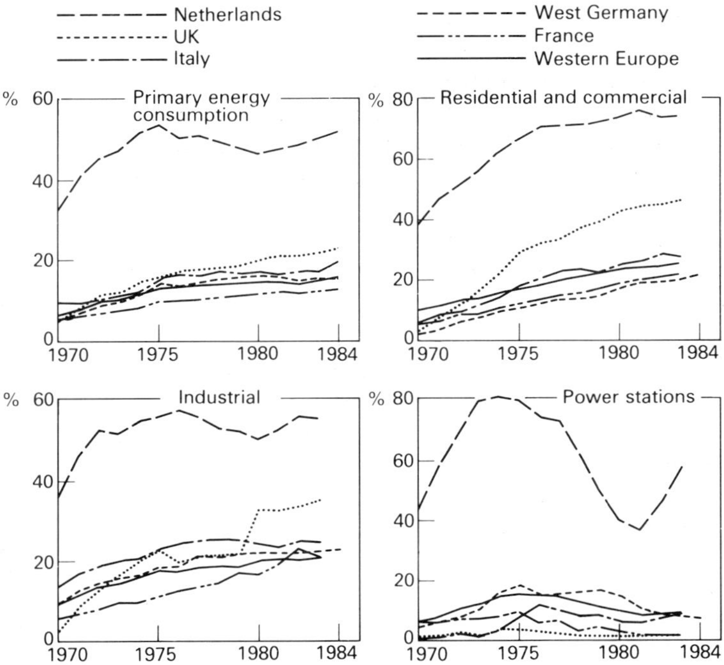

Figure 3.4 Natural Gas Market Penetration in Western Europe, by Sector.

declined by 17 per cent between 1979 and 1983 and the gas share in power generation dropped from 8 to 6 per cent.

The Significance of Growth since 1982–3

Compared with this somewhat gloomy period from 1979 to 1982–3, recent energy demand trends may foster some optimism and indicate a return to an age of growth. However, short-term developments are easily mistaken for long-term trends. Table 3.4 shows that West European primary energy consumption rose by a total of 4 per cent between 1982 and 1984. It is worth noting that gas, for which a growth rate of 11 per cent was recorded, was more successful than other sources of energy. However, some qualification is required, mainly regarding power station fuel consumption and the influence of weather conditions on the space heating market. In the

Table 3.4: Energy Consumption. 1982 and 1984. Million Tonnes of Oil Equivalent.

	1982	*1984*	*Change 1982/1984 %*
Western Europe:			
Primary energy consumption	1,159	1,205	4.0
Final energy consumption	780	790	1.3
Gas consumption	171	190	11.1
West Germany			
Primary energy consumption	253.1	263.6	4.1
Final energy consumption	164.5	171.2	4.1
Gas consumption	38.0	41.2	8.4
Temperature-adjusted gas consumption	39.0	41.0	5.1

Source: Ruhrgas

Netherlands, for instance, a quantity of 20 bcm has been sold, for a period of five years, to electric utilities at a reduced price. In Italy, power stations will absorb additional gas – if competitive with heavy fuel oil (and imported electricity) – for a number of years to come, as more cost-efficient base-load nuclear power generating capacities and less costly medium-load coal-fired power plants have not been built. In Belgium, gas has been able to oust fuel oil temporarily as fuel oil prices have risen sharply. If extra gas sales to power stations in these countries are deducted, the overall growth rate between 1982 and 1984 is reduced from 11 per cent to approximately 7 per cent. If an adjustment is also made for weather conditions, growth rates become very modest indeed, as the West German example shows.

Although these figures do not justify any high-flying hopes, it was possible to halt the downward trend in the gas market in 1983 and moderate growth has returned. This turn in fortunes was caused by an economic recovery (there was a 2.3 per cent increase in GDP in 1984) and was at the same time a product of an adjustment of gas prices to the new market environment agreed by major European gas companies with producers inside and outside Western Europe under the regular price review clauses now incorporated into most international gas contracts. The adjusted prices are still largely based on oil products prices, but establish a stronger linkage with gas oil prices and reflect stiff inter-fuel competition, especially for steam-raising outlets.

Trends in Energy Demand

When extrapolating future energy demand trends, it must be accepted that Western Europe is too heterogeneous a region to justify an assumption of parallelism of energy demand parameters in the different countries. However, certain conclusions can be derived for Western Europe from data for West Germany, first because this country is the largest energy consumer in the region, with an essentially deregulated energy market, and secondly because trends in this nation would appear typical of overall tendencies in Western Europe.

Changes in the parameters for energy consumption in West Germany since 1973 are shown in Table 3.5.

Table 3.5: Energy Consumption in West Germany. Percentage Changes from 1973 to 1985.

GDP (in real terms)	+21
Industrial output	+ 7
Total car mileage	+34
Total floor area of dwellings	+29
Primary energy consumption	− 0.5
Final energy consumption	− 3.7
Energy ratio	−18

Against this background, energy demand forecasts have altered radically since 1973 (see Figure 3.5). Critics often query the mood of moderation that now prevails and ask whether the trends of recent years have not been exaggerated. At times, they are moved to voice the suspicion that predictions reflect deliberate pessimism in order to promote a buyers' market climate. Admittedly, plans for the future are all too often guided by present-day trends in all spheres of human life, including, of course, the energy industry, and the tyranny of the status quo frequently rules human thought. Regarding the future development of energy consumption, however, it is essential to recognise and appreciate the three elements that will play a key role. These are overall economic development, structural changes in industry where energy-intensive sectors will continue their decline and, last but not least, the sustained energy conservation efforts.

Economic growth is certainly a factor that will have an important effect on future energy consumption. Current forecasts in West Germany assume an average annual growth rate of 2.0–2.5 per cent in real terms until the year 2000. These assumptions do not seem unduly pessimistic, since the base for the high rates of growth during the period of reconstruction was low and the absolute expansion now predicted is certainly comparable. Most energy demand forecasts made during the early 1970s were based on higher growth rates, but even during the period 1970–84 – when two series of oil price rises left their mark on economic activity – the average annual growth rate was 2.2 per cent (but only 0.7 per cent during the first half of the 1980s).

The economic growth rate predicted by analysts for the period from 1985 to the year 2000 does not really reflect a new atmosphere of pessimism but, rather, a scenario of medium-term and long-term progress:

(a) Indispensable structural changes in industry will be furthered.
(b) Protectionism and trade restrictions will not extend any further.
(c) The world financial system will not collapse.

On the other hand, certain developments limiting growth must not be overlooked:

(a) The population is declining.
(b) Private sector consumption is nearing saturation point, as is demand for new private homes.
(c) Government spending on public works such as new roads and hospitals will tend to decline.

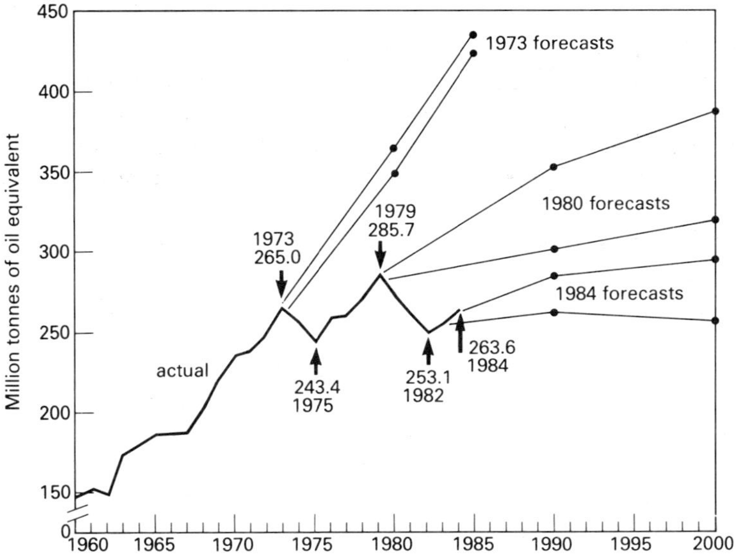

Figure 3.5 Primary Energy Demand Predictions for West Germany.

(d) In spite of high productivity in West German industry, the high level of salaries and wages and, above all, high fringe benefits threaten the international competitiveness of manufacturers based in West Germany, which earns some 30 per cent of its GNP abroad – more than any other country.

Though no detailed reasoning is put forward at this point, this picture of the future still holds true when higher past, present and possibly future growth rates in the USA and Japan are referred to by more optimistic outside observers (see Figure 3.6).

Translated into energy demand trends for West Germany, these general economic prospects indicate hardly any growth in the various sectors of the energy market. Regarding the future development of residential and commercial energy consumption, it is important to realize that the number of dwellings is unlikely to continue to increase and dwelling sizes also seem to have peaked. In addition, energy conservation will continue, mainly as a result of the gradual replacement of obsolete space heating systems by considerably more efficient

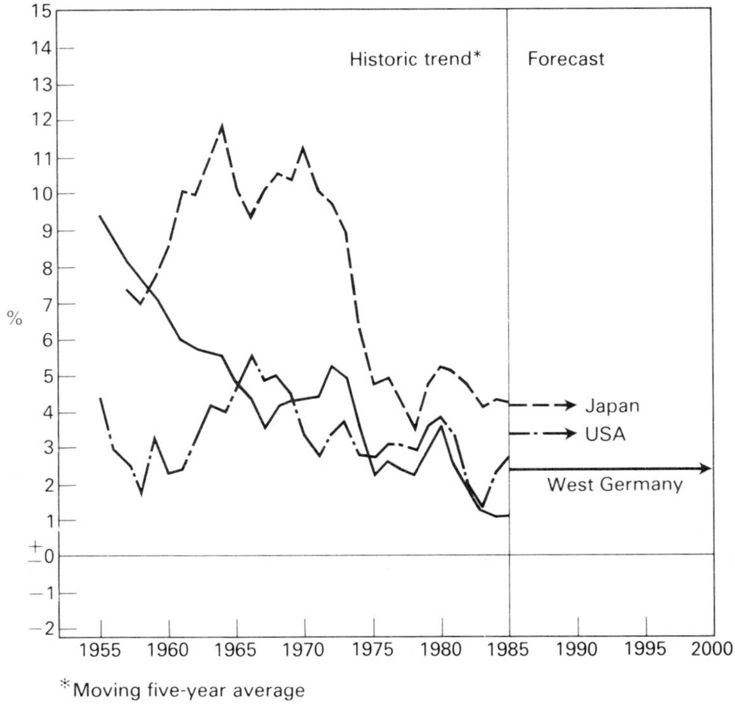

Figure 3.6 Historic and Predicted Trends of Economic Growth in West Germany, the USA and Japan.

modern plant and improved thermal insulation, rather than because of changes in consumer behaviour slowing down the growth of residential and commercial energy demand in the wake of energy price rises.

More than 3 million boilers will have to be replaced in West Germany during the course of the next ten years. The average efficiency over the year of modern boiler plants equipped with advanced electronic controls may be as much as 85 per cent, reducing energy consumption for space heating by approximately 25–30 per cent. On housing estates, small district heating systems or co-generation plants could be as much as 35 per cent more efficient than equivalent modern boiler installations.

Industrial energy consumption has not risen since the second half of the 1960s if the effects of limited economic booms in certain years are disregarded. The impact of energy conservation is reinforced by structural changes, as energy-intensive branches of industry are left behind by other sectors. During the period 1979–84, industrial energy consumption in West Germany declined by 15 per cent, and the structural changes referred to account for about 40 per cent of the decrease. The energy requirements of Hoechst AG, the West German chemical group, are but one example. Hoechst AG has increased its output by some 50 per cent since 1975, but energy consumption has remained unchanged. For the transport sector, no further growth is predicted and fuel consumption may even decline.

With this background in mind, a primary energy demand increase of a mere 6 per cent by the year 2000 has been predicted for West Germany. This rise translates into a very modest annual growth rate of 0.4 per cent. The degree of agreement between industry, economic institutions and Government consultants on this assessment of future trends is remarkable.

For Western Europe as a whole, a somewhat faster growth of 1 per cent per annum has been predicted, as industrialization progresses in certain nations. At this rate, energy consumption would rise from some 1,200 mtoe in 1984 to 1,400 mtoe by the year 2000.

Trends in Gas Demand

The prospects for natural gas must now be evaluated in the context of the overall energy environment discussed. In the residential and commercial sector, the outlook for gas is encouraging. One in every two new West German dwellings is today heated by gas. However, the sensitivity of this market sector to changes in prices and costs is high, as the French example has shown. In France, the national electricity company responded to its surplus of nuclear generating capacity by taking almost two-thirds of the space heating market in the new dwelling sector on the strength of low electricity prices, and reduced the gas industry's share from over 28 per cent in 1980 to some 20 per cent within a period of three years.

The main competition faced by gas in the residential and commercial sector is currently represented by gas oil. Gas oil sales now seem less affected by the general concern about the security of supplies and price rises, but increased awareness of the environmental issue tends to favour gas. Even so, the gas industry must not focus its pricing policy in this sector exclusively on gas oil price trends at a time when the market shares of electricity and district heating are increasing continually. The important question will be whether gas oil remains the marker fuel in inter-fuel competition, and this may become apparent if gas oil loses a major part of its market. Until recently, the fuel defended its market share successfully and only lost ground during the last few years when it held a smaller percentage of the new dwelling market than of the space heating market as a whole.

The marketing efforts of the gas industry are not, however, limited to arguing with gas producers for market-oriented pricing and offering competitive prices to end-users. The industry also strengthens the position of gas by major activities of its own, including:

(a) The development of even more efficient, less polluting and more easily controlled gas appliances.
(b) The provision of consumer consulting services and advertising.
(c) The development of new marketing strategies, such as the supply of space heating and domestic hot water by local utilities to end-users by contracts under which the utilities will also operate space heating installations.

As building activity is declining, the replacement market and commercial outlets must remain the main targets of gas sales efforts. The West German gas industry anticipates that it will be able to win an additional 1.5 million residential customers during the latter half of the 1980s, thereby increasing the number of homes using gas for space heating by some 20 per cent.

The industrial sector of the energy market is highly complex. Throughout Western Europe, the oil substitution potential in this sector is lower than in the residential and commercial sector (cf. Table 3.2). In West Germany, for instance, the share of 17 per cent presently held by oil is less than half the share of oil products in the residential and commercial market (42 per cent). Nevertheless, prospects for gas are encouraging, particularly as regards small and medium-sized outlets. Stricter legal regulations for small plants, as in West Germany, tend to favour the use of gas in cases where gas meets the environmental standards without flue gas treatment.

In the steam-raising market, coal is the main competitor and the progress of fluidized-bed combustion techniques will have to show whether the competitiveness of coal is further enhanced by a reduction in cost and improved load control techniques. In process heat applications, electricity has already made major inroads. Presently, it covers 9–10 per cent of energy demand for these applications in West Germany. The relative movements of electricity and gas prices will determine the extent to which gas will be able to meet this demand in the future. For the gas industry it will be essential to strengthen its position by further improving the efficiencies of gas-fired industrial plant and by developing advanced industrial systems for integration into automated manufacturing lines.

In the feedstock market, representing today 9 mtoe or some 5 per cent of gas consumption, operators may eventually move to countries where the cost of providing energy is considerably lower than in Western Europe.

Regarding the generation of electricity for public distribution, the long-term development of gas sales will remain a political subject in Italy and in the Netherlands. In West Germany, following a drop in gas sales to electric utilities by half, the gas industry is set to defend its remaining market share by competitive pricing, especially for sales for the combined generation of electricity and heat at power stations or industrial sites near to cities. In the UK and France, on the other hand, gas is hardly used for the production of electric power.

Considering existing surplus power-generating capacities and the cost of peak-shaving associated with the use of gas for the generation of peak electricity, it would appear doubtful whether progress in gas turbine engineering could give gas a larger share in the West European power station market. The prospects for gas turbines may be better in base-load industrial co-generation applications.

More stringent environmental legislation will certainly tend to favour natural gas over other fossil fuels. As a result of new emission requirements, differences between the emissions produced by the various fuels are becoming smaller, but in most cases, natural gas must still meet higher standards, making it, as in the past, a superior fuel in terms of environmental protection.

Security of supplies, another consideration which may tip the balance for or against a given source of energy, may be one of the less prominent criteria of decision-making processes today because of the current temporary energy surplus. However, it will return to the limelight if and when the share of growing world oil demand covered by OPEC oil increases from its current level of approximately 40 per cent to between 50 and 55 per cent (excluding demand in centrally planned economies) in the year 2000, as recently forecast by the major international oil companies.

Whereas the trend of gas sales to residential and commercial users remains steady, a number of uncertainties make it much more difficult to quantify the long-term prospects for gas in the industrial market. If, in spite of these uncertainties, an attempt is made to predict overall future gas sales, a 10–15 per cent increase in gas supplies to West German consumers by the year 2000 would seem a reasonable estimate. This rise would give gas a share of 16–17 per cent in the primary energy market compared with 15.6 per cent in 1984.

For Western Europe as a whole, gas consumption is anticipated to increase from the current level of 190 mtoe to 220 mtoe or, put in different terms, 290 bcm by the year 2000.

The Potential for Growth Management

With natural gas production currently running well below the existing production potential, growing gas reserves both inside and outside Europe, and an increased awareness of environmental issues, gas producers, the EEC Commission and the IEA have raised the question of an expansion of gas supplies beyond current target sales. The implications of such a change in market patterns would be different in each West European country and must be evaluated in the light of national economic and political conditions. The analysis in this paper will be restricted to West Germany. It is, of course, true to say that a major increase in gas supplies would not be prevented by technical limitations, as the substantially higher natural gas penetration rates in the USA, the USSR, the Netherlands and the UK show (see Figure 3.7).

Gas sales could, however, only be stepped up if more gas were substituted for other sources of energy. Straightforward economic reasons prevent gas from penetrating the market held by nuclear energy and lignite, as the costs associated with these two sources of energy and, above all, the marginal costs are too low. Outlets for West German hard coal are protected by Government policy reinforced by long-term agreements. The delicate nature of competition with coal had to be recognised by the West German gas industry even at a time when it was losing markets to coal, but was nevertheless under public pressure for not leaving a sufficiently large market to West German coal.

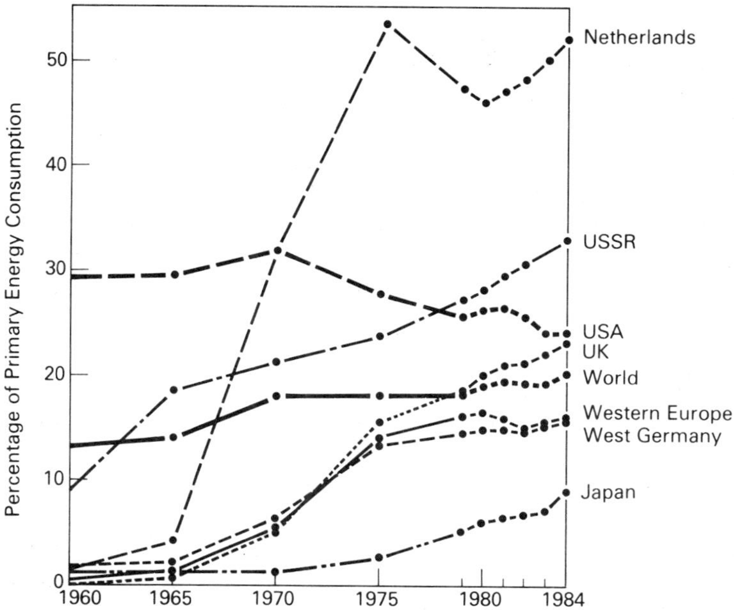

Figure 3.7 The Share of Natural Gas in World Energy Markets.

The prime target for substitution by gas would therefore have to be gas oil in the residential and commercial sector and heavy fuel oil and gas oil in the industrial sector. To accelerate the substitution process, a substantial reduction in the price of gas would be necessary. In the residential and commercial market it would be impossible, for commercial reasons and because of the restrictions imposed by antitrust legislation, to introduce price discrimination between the existing customers of the gas industry and potential new clients. It would therefore be necessary to introduce lower rates for all customers, but it is hard to imagine that gas producers not supplying additional gas for an expansion of the gas market would be willing to reduce prices under existing supply contracts in order to improve the substitution potential from which other producers would benefit.

In the industrial sector somewhat different considerations apply, since it would be possible to allocate new outlets or outlets defended by isolated major price reductions to sellers willing to market their gas at a lower price. However, the potential for action is again limited for a number of reasons, not least that the main target of a more aggressive marketing policy would have to be oil, which holds only a 17 per cent share in this market sector. Further, such additional gas sales would only be an attractive proposition to producers if no higher-priced outlets could be found for the gas, even in the longer term, and if the reduced price still produced acceptable returns.

From a general political point of view, it is also necessary to ask whether an over-aggressive gas marketing strategy would not eventually jeopardize the long-term objective of the diversification of energy supplies in the country. Seen in this light, it has certainly been a wise policy of the West German gas industry to declare as long ago as the early 1970s that it would strive only for a limited market share. The industry considers that a share of some 18 per cent in the primary energy market is a reasonable target and that it would be wrong to call this target figure into question today.

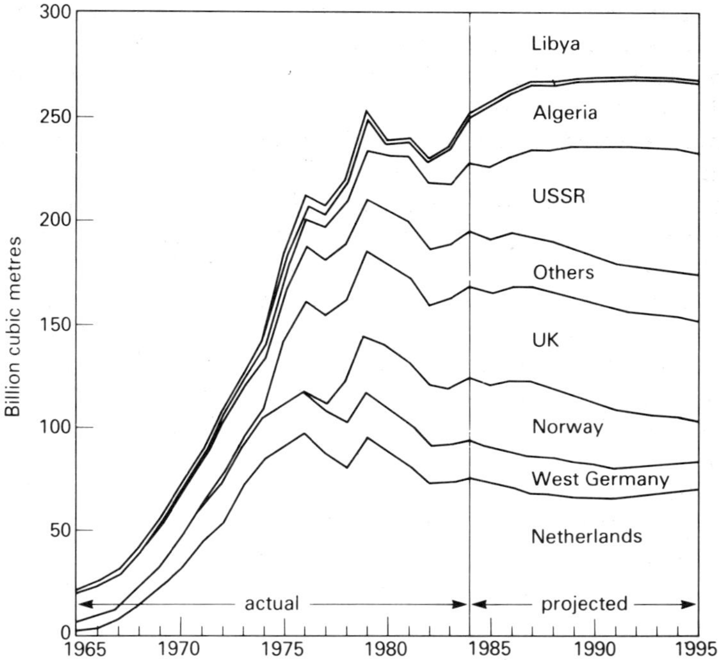

Figure 3.8 Natural Gas Supplies to Western Europe.

The Sources of Future Gas Supplies

The gas demand development discussed and the targets of the gas industry described will, of course, represent a realistic view of the industry's future only so long as governments do not intervene and adequate, secure and reliable supplies are available or become so under competitive conditions.

Figure 3.8 reviews gas supplies to Western Europe and shows foreseeable developments until 1995 on the basis of agreed import contracts and predicted indigenous production. A total of 23 per cent of all West European gas supplies was provided by external sources in 1984, of which the USSR alone accounted for 14.5 per cent and Algeria for 7.8 per cent. Indigenous production is mainly attributable to fields in the Netherlands, the UK and Norway which together held a share of 59 per cent of West European gas supplies in 1984.

If, for the period until the mid-1990s, it is assumed that the industry will see a return to higher load factors under operational contracts, an expansion of gas imports from Algeria and the USSR on the agreed scale, and a development of indigenous gas output in line with current expectations, then gas supplies will be sufficient to cover estimated West European demand until the mid-1990s. A need for major new cross-border gas supplies will not arise until the latter half of the 1990s when supplies under operational contracts will decline and gas will be needed to serve a somewhat larger market. New cross-border projects supplying some 33 bcm per annum may then be necessary by the year 2000 (see Table 3.6).

Table 3.6: Trends in Supply and Demand of Natural Gas in Western Europe. Billion Cubic Metres.

	1984	*1990*	*2000*
Domestic supplies	129	118	124
Intra-trade	65	62	56
Assumed new Norwegian project (Troll)			(18)
Contracts for imports from outside the region (including renewals)	58	85	110
Potential for further import projects			(15)
Total	252	265	290

Note: Data for 1984 are tentative.
Source: Ruhrgas

Substantial exploration and development efforts will be necessary, particularly in the UK, Italy and West Germany, to warrant the anticipated long-term contribution from the domestic supply base. The Netherlands and Norway, discussed below, will remain particularly important countries for the West European gas industry, since their exports to other West European nations present important international streams of gas in the West European pattern of supplies.

In the *Netherlands*, additions to domestic reserves and the more moderate domestic demand forecasts were translated into political and commercial action. Existing export agreements were renewed for periods of ten years and will in most cases now terminate after the year 2005. Annual sales of some 26 bcm provided for under the renewed contracts are equivalent to some 10 per cent of West European gas demand. The implications of the Dutch decision thus match the impact of the 1981–2 Soviet natural gas export project. The two projects taken together have altered the West European gas market. They have, on the one hand, given firm long-term market shares to the two producers and, on the other hand, put West European consumers at ease, now supplied with adequate quantities of gas for the next ten years and only leaving room for limited additional imports during the remainder of this century. The potential for imports of gas to West European nations, both from within the region and from outside, has shrunk to scarcely more than one-third of its original size in recent years.

The question of additional exports to the UK or a further renewal of sales to other countries on the Continent will, in the final analysis, be a political issue in the Netherlands. The development of domestic gas reserves, trends in Dutch gas consumption, budget requirements and the status of nuclear energy will be elements that will have to be considered at the time when this issue is to be decided.

Norway, boasting gas reserves totalling 2,700 bcm, equivalent to 44 per cent of all West European reserves, covers only 12 per cent of West European gas demand. This relatively small share is, of course, explained by the recent nature of major additions to Norwegian reserves and the long lead times required for the development of the technology and the installation of the systems needed to exploit these reserves.

However, the pattern of supplies from Norwegian waters during the last few and the next ten years shows that Norway has failed to ensure continuity in the country's gas exports (see Figure 3.9). Field development during this period has a patchwork appearance, with corporate decisions made field by field determining the dimensions of each patch. In spite of Norway's excellent resources, the share of Norwegian gas in the West European market will decline well into the 1990s, since other producers have already seized the market opportunities available until that time.

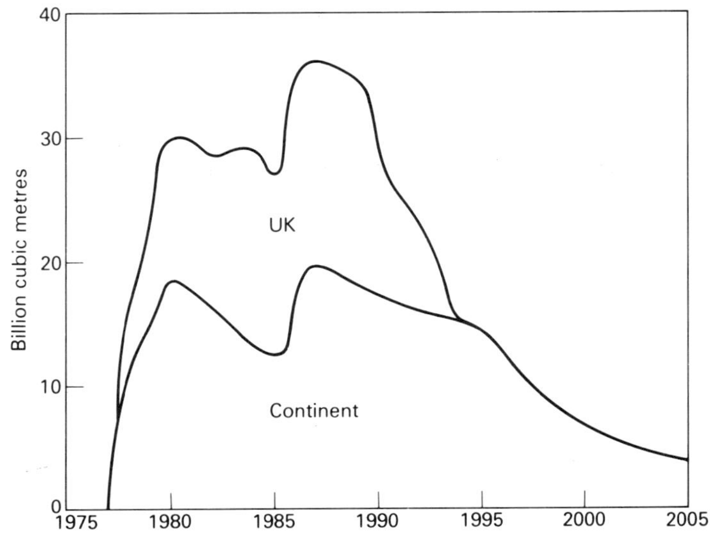

Figure 3.9 Norwegian North Sea Gas Supplies to Western Europe.

The Troll field development now marks a crucial turning point for Norwegian gas exports, allowing the country to replace the patchwork by a tailor-made apparel suited to the needs of a stable long-term gas export strategy and to maintain or perhaps even to increase slightly the nation's share in the West European market. The Troll field will enable Norway to move towards a supply-oriented marketing policy under which definite quantities of gas would be sold for definite contract periods, leaving the decisions as to which fields provide exports and when to internal Norwegian decision-making processes. The Troll field would provide a firm basis for such a policy, since it allows considerable flexibility in field development and operation without the restrictions implicit in reservoir depletion uncertainties.

Such deliberations seem to be well under way. Negotiations between Norwegian producers and a group of buyers from Continental Europe, including Ruhrgas AG, envisaging the export of major quantities of gas from the mid-1990s onwards, have already commenced and, although the major hurdles have still to be faced, these are making progress. The key question that remains to be answered is whether the market price of energy in Continental Europe provides a sufficient economic driving force for the project. In the overall economic analysis, the take of the Norwegian Government, estimated to total more than 50 per cent of the cost of gas at beach, is a decisive factor and the future will show whether the producers consider that the feasibility of the project hinges upon the issue of the Government take. If they come to this conclusion, the matter will have to be settled between the producers and the Norwegian Government. The buyers have no choice but to offer the market price. They are not subject to comparable Government take requirements. In sum total, however, the first phase Troll project currently under negotiation would seem to fit the needs of the gas market in Continental Europe.

The *USSR* is Western Europe's largest external gas supplier, representing Europe's second largest supply base after the Netherlands. Deliveries from the USSR, totalling some 31 bcm (as delivered), accounted for 14.5 per cent of West European gas supplies in 1984. Contracts agreed with the Soviet exporter call for an increase in annual deliveries to between 45 and 50 bcm by 1990 (see Table 3.7).

Soviet natural gas reserves exceed the 40,000 bcm mark and represent 42 per cent of world gas reserves. Although most gas fields are located in a hostile environment in the vicinity of or

Table 3.7: Soviet Natural Gas Production and Exports. Billion Cubic Metres.

	1980	*1983*	*1984*	*1990*	*1995*
Domestic consumption	381	477	514	730	880
Exports to centrally planned economies	31	34	42	50	55
LNG exports to Japan	0	0	0	0	5
Exports to Western Europe	23	25	31	50	60
Total	435	536	587	830	1,000

Source: Ruhrgas

even above the Arctic Circle in West Siberia, the construction of six 56-inch pipelines from the Urengoi reservoir to the West of the country and the increase in West Siberian gas output to more than 1 bcm/d under the 1981–5 five year plan have demonstrated that the Soviet Union has the technology required to mobilize these reserves. According to the targets set by the 1986–90 five year plan, annual Soviet natural gas production will be stepped up to more than 800 bcm and an annual Soviet gas output of 1,000 bcm in the mid-1990s would not appear unlikely. It will be necessary, however, to develop fields in the far North and the technical challenges will become greater. The first step in this direction will be the development within the next five years of the Yamburg field, situated some 150 km north of the Urengoi reservoir.

Agreed gas exports to Western Europe will account for only 6 per cent of Soviet gas production in 1990. The resource base and the technical capabilities of the country would therefore allow a further expansion of sales to Western Europe without any great difficulty. As, on the other hand, gas demand forecasts for Western Europe are moderate, alternative gas supplies are available and the West European gas industry is pursuing a policy of diversification, no major new exports to countries already receiving large quantities of Soviet gas are at present anticipated. Minor projects for exports to other West European nations seem to be in the offing.

The purchase of Soviet gas by the UK would be primarily a political question. However, even if the political answer were in favour of such a project, it would remain to be seen whether the Soviet exporter would be willing to supply gas at a price which would be competitive at the point of arrival in the UK, considering the cost of transporting the gas over approximately 1,000 km in Western Europe. Perhaps the question of Soviet gas imports to the UK has only become an issue of public debate because certain groups hope that a cross-channel pipeline would be a spin-off effect of this debate. Those involved in the discussion now seem to appreciate that such a pipeline would only be feasible if it were supported by a long-term supply agreement guaranteeing its amortisation.

As regards the frequently discussed dumping offers by the Soviets, the West European gas industry does not see any reason for such a Soviet policy and, even if such offers were made, they would not change the gas industry's policy of diversification of supplies.

The LNG Option. Liquefied natural gas is another source of West European gas supplies. LNG deliveries of 14 bcm in 1984 accounted for 6 per cent of the gas serving the West European market. The major LNG supplier is Algeria which provided 13 bcm in 1984. Its huge reserves of over 3,000 bcm and its proximity to the West European market make Algeria a strong candidate for new West European imports. However, it apparently missed the opportunity to establish itself even more firmly in the West European market when the decision was taken in 1980 not to go ahead with the West German/Dutch projects calling for annual sales of 24 bcm. Algeria will now find it difficult to compete with other suppliers for the limited outlets in Western Europe in the second half of the 1990s and beyond, or to sell more than minor volumes of gas to southern Europe.

The Algerians have a vested interest in early LNG sales in order to gain trade for their idle liquefaction plants and LNG carriers, but it will be necessary not only to adapt their pricing policy to the new market environment but also to convince potential buyers that reliable LNG deliveries will be guaranteed over a reasonable contract term commencing in the mid-1990s if the contract is served by the liquefaction plants which will by then be fifteen years old. This latter restriction could be circumnavigated if new Algerian gas exports were handled by the Transmediterranean pipeline, though new facilities would be required for such a project.

In spite of spare capacity at existing terminals and terminals under construction, it would appear doubtful whether this century will see a major LNG project implemented for the importation of LNG to Europe from other parts of the world. West European gas demand is limited, the cost of gas liquefaction has risen sharply, and potential suppliers such as Nigeria or perhaps even Qatar would have to supply the gas over long distances.

A Buyers' Market and an End to Oil Price Pegging?

At a time marked by predictions of moderate growth of demand and a stated interest in the sale of gas on the international market expressed by suppliers in the USSR, Norway, Algeria and other countries, some argue that the international gas industry has entered a phase of trading reflecting a buyers' market and allowing the 'unpegging' of the gas price from the price levels of competing fuels such as fuel oil.

It is clear that producers planning to sell new gas for delivery in the next five or ten years will be faced with a buyers' market, if this is simply taken to mean that supplies exceed demand. Under these conditions, it may be possible to purchase new gas at a lower price, but the price would have to be sufficiently low to generate new demand. Furthermore, the price of gas supplied under long-term agreements is determined by inter-fuel competition at the end-user's burner-tip rather than supply and demand laws applicable to small extra volumes of gas.

The long-term market for gas delivered after the mid-1990s is even less of a buyers' market. Supply and demand would, on the contrary, seem to be in equilibrium, if, in the balance between supply and demand, supplies are restricted to reliable projects viable from today's viewpoint.

Apart from these supply and demand considerations, the very principles of gas trading do not allow gas pricing without making reference to the prices of competing sources of energy. The gas trade is a highly capital-intensive business and gas plant and equipment cannot readily be used for any operation other than the project for which they have been built and installed. For this reason alone, the long-term nature of the investment must be matched by an equivalent period of amortisation. It is therefore necessary to agree a pricing mechanism which guarantees that gas will always be competitively priced for the end-user, since volume, the other factor in the formula of supply and demand, cannot serve as a corrective element as it does in the oil industry. Gas pricing systems must therefore always take their orientation from the prices of competitors. However, it cannot be denied that the margin earned by the gas industry between the field and the point of delivery to the end-user is influenced, to a certain degree, by supply and demand. Furthermore, oil products will not always or exclusively be the marker fuels. The recent past has witnessed not only a partial replacement of heavy fuel oil price pegging by gas oil price pegging but also an extension of price formulae to account for competition from coal in the steam-raising market. The price of electricity may become an additional reference point for gas pricing.

Summary

The gas industry has been unable to escape the woes of the oil price rises in 1979–80. In a market in which sales have declined, it must accept sharply reduced growth prospects for the future. The tools of the industry have proved adequate to handle these radical changes. Prices have adjusted to the new competitive environment and the industry has not been forced to abandon major plants and facilities as was the case in the oil industry, particularly in its refinery operations.

Predicted indigenous production levels and supplies agreed under import contracts seem adequate for medium-term demand, and lead times are comfortable for contracting new gas to cater for the needs of the second half of the 1990s and beyond. Prospects are encouraging for the timely conclusion of such contracts from reliable sources.

Gas has remained a source of energy whose assets are appreciated both by governments and by consumers. It makes an important contribution to the protection of the environment and helps to diversify the sources of energy supplies. Competitive pricing and reliability remain the *sine qua non* for retaining this goodwill. Even under these conditions, however, the scope for an expansion of gas sales is very limited indeed.

4 NORWEGIAN GAS: RESERVES, PRODUCTION AND MARKETS*

Arve Johnsen

Introduction

In 1984 the most important gas fields on the Norwegian continental shelf were Ekofisk, Sleipner, Heimdal, Frigg, Oseberg, Troll, Gullfaks, Statfjord, Tyrihans and Askeladden (from south to north). Some of these fields have associated gas (Ekofisk and Statfjord), while some have gas only (Sleipner and Frigg). Some have rich and some dry gas, and in addition to the variety in quality there is also a broad geographic spread. The distance between Ekofisk in the South and Askeladden in the North is close to 2,000 km (see Figure 4.1).

Reserves

According to Statoil's estimates, proven, probable and possible petroleum reserves south of Stad are about 5–6,000 mtoe, and approximately half of these reserves are estimated to be gas. North of Stad the reserve situation is less clear. However, considering the size of the continental shelf and the drilling results so far, reserves could be up to three times as high as those south of Stad, and the gas share may turn out to be similar.

Proven Norwegian gas reserves are around 2,500 mtoe which corresponds to around 2 per cent of the world's natural gas reserves and approximately 35 per cent of the natural gas reserves in Western Europe. Until now gas reserves have been developed only in water up to 200 metres deep. However, more than 60 per cent of the probable and possible gas reserves south of Stad are located in water more than 300 metres deep. North of Stad it is likely that an even higher percentage of probable and possible gas reserves are located in deep water.

* Paper presented at the Oxford Energy Seminar, 1984.

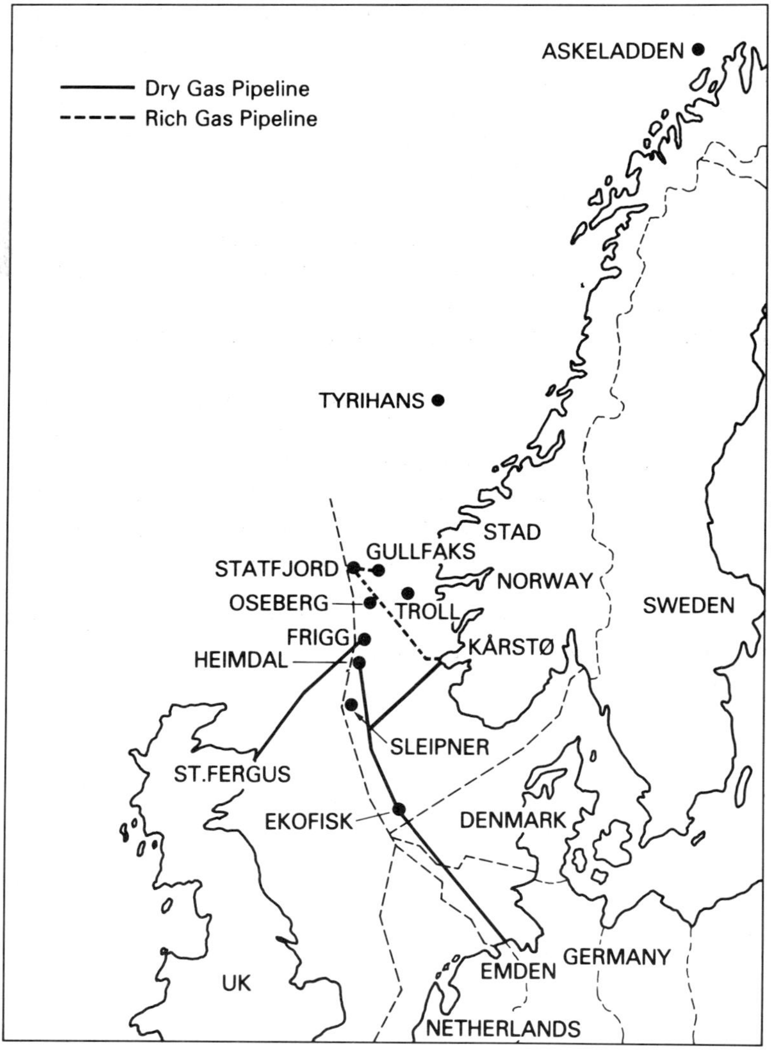

Figure 4.1 Norwegian Gas Provinces and Transportation System.

Production

Present Norwegian gas production comes from the Ekofisk and Frigg areas, but during the latter part of this decade new gas production will be phased in from Statfjord, Gullfaks and Heimdal. The production of gas from the Ekofisk and Frigg areas now represents approximately 25 mtoe per annum, and by the end of this decade annual production will reach around 30 mtoe. It remains to be seen what will happen to Norwegian gas production in the 1990s and after the year 2000 – production could range from 30 to 50 mtoe per annum – but it is clear that price will be the determining factor.

Transportation

Transportation systems for gas have been built from Ekofisk to Emden in West Germany – Norpipe – and from Frigg to St. Fergus in Scotland. A new gathering line for rich gas from Statfjord and Gullfaks via Kårstø in Norway to Ekofisk is under construction, and will be in operation in 1986. The system will also transport the dry gas from the Heimdal field. The overall length of the Statpipe system will be around 880 km. The rich gas line from Statfjord to Kårstø will have a capacity of 8 bcm per annum, and the rich gas landed at Kårstø will be treated at a terminal where the NGLs will be extracted. A dry gas line with a capacity of 6 bcm per annum will be built from Kårstø to a riser platform located east of the Sleipner area, and a separate dry gas line from Heimdal to the same riser platform, including a connecting 36-inch line between the platform and Ekofisk which will complete the system.

Landing a pipeline on the Norwegian coast is difficult, both because of the rough sea-bed and the wave forces. In the case of Statpipe a 600-metre long prefabricated tunnel was installed from the land into the sea, where it was placed on pre-cast foundations on the sea-bed. The tunnel was filled with water after it was positioned, and the two pipelines – one incoming and one outgoing – were pulled through the tunnel. Thereafter, pipelaying continued into deeper water and across the Norwegian trench. On land the pipeline from Kalstø to Kårstø has been laid partly in a trench and partly in tunnels under various fjords. The total Statpipe project, including pipelaying, riser platforms and the terminal, is on schedule and within budget. Both the Statpipe and Norpipe systems have flexibility to allow for new gas fields in the future.

The total Norwegian transport capacity is about 20 bcm to the European Continent and about 15 bcm to the UK (approximately 30 mtoe altogether). Based on gas production from existing fields, utilisation of capacity in the Statpipe/Norpipe transportation system will be between 80 and 90 per cent until the beginning of the 1990s. Thereafter capacity utilisation will fall. In the case of the Frigg line, 100 per cent of its capacity will be available as from the middle of the 1990s. Both the Statpipe/Norpipe transportation system and the Frigg line allow for additional deliveries of gas in the 1990s from the Norwegian continental shelf to Continental Europe, as well as to the UK.

Markets

The major gas-consuming areas in the world are the USA, the USSR and Western Europe. In the period 1970–80 total world consumption of gas increased by 3 per cent per annum, whereas in Western Europe the increase was close to 10 per cent per annum. Norwegian natural gas is currently exported to Continental Europe and the UK.[1] These two areas are considered the most natural markets for future Norwegian gas sales, although Scandinavia and the USA may become potential markets.

Continental Europe. Consumption of natural gas in Continental Europe was between 140 and 150 mtoe in 1980 and 1981. In spite of the reduced natural gas consumption experienced during 1982, demand for natural gas recovered in 1983 and is likely to increase throughout the 1980s and 1990s. However, projections for supply of and demand for natural gas in the Continental European market towards the turn of the century indicate changes in the structure of the market. The most significant changes may be a reduction in the indigenous supply and the consequent need for new imports. Nevertheless, the

[1] 'Continental Europe' is defined here as Western Europe (the OECD countries) less Scandinavia and the British Isles.

most important indigenous source of supply will still be Dutch gas. Although Dutch production may fall slightly towards the year 2000, exports from the Netherlands will still be a major element in Continental supplies for the rest of this century. By the middle of the 1990s the French Lacq field will be depleted and West German and Italian production will be reduced.

The present level of imports from the USSR, Algeria and Norway will be sufficient to cover the remaining demand for natural gas, at least until 1990, but thereafter Continental Europe will become increasingly dependent on new imports of gas. The suppliers of these new imports will be the USSR, Algeria and Norway, with possibly some producers in Africa and the Middle East supplying LNG. However, the USSR is likely to be the major supplier of new imports to Continental Europe for the following reasons.

First, the USSR already has a key role in supplying gas to Continental Europe. Secondly, with the new pipeline from Siberia to Western Europe, the USSR will have a total transportation capacity for natural gas of around 60 mtoe per annum and contracts concluded to date will utilise approximately 70 per cent of the system. Consequently, the capacity will exist for increased Soviet deliveries in the 1990s without the need for new pipelines. Thirdly, the USSR holds approximately one-third of the natural gas reserves in the world and will in all probability sell gas at competitive prices, whether by comparison with other gas suppliers or other sources of energy like oil, coal or nuclear power. The fact that gas exports are likely to be one of the best foreign currency earners for the USSR further explains this sales policy.

Algeria is also an important gas supplier and, through the Transmed pipeline, Algerian gas sales to Continental Europe could reach 30 mtoe in 1990. In addition, boosting could increase this capacity in the 1990s. It can be seen, therefore, that new Norwegian gas will never be an alternative to Soviet and Algerian supplies, although Norwegian supplies will constitute an important supplement to these sources. To what extent new Norwegian gas will be bought by Continental European importers cannot be predicted, but there is no doubt that the market is of interest to Norwegian exporters.

The UK. In 1982 and 1983 consumption of natural gas in the UK was between 41 and 45 mtoe, and demand is likely to increase throughout the 1980s and 1990s. The UK natural gas market is characterized by a high degree of self-sufficiency and by the fact that Norway will be the main foreign supplier of gas in the 1980s. Contracted indigenous production will decrease markedly towards the end of this century, mainly because of reduced production from the fields in the Southern Basin. Probable and possible new indigenous production will not be sufficient to prevent an increasing gap between domestic production and increased demand from 1990 onwards. In addition, Norwegian exports of gas from the Frigg area will cease by the mid-1990s. The potential suppliers of the necessary new imports in the 1990s are the Netherlands, Norway, Algeria, West Africa and perhaps others. Given the favourable location of Norwegian gas reserves and the incentive to utilise present pipeline capacity in the North Sea, there may be imports from new Norwegian fields in the 1990s.

Scandinavia. There is as yet no natural gas market in Scandinavia, although Denmark is in the process of establishing a gas transmission and distribution system. This is expected to be operative from October 1984 and will utilise gas from the Danish continental shelf. Sweden also has a distribution system under construction and agreement has been reached with Denmark for imports of gas from 1985. In Norway there is not yet any distribution system for natural gas, whereas Finland imports approximately 1 bcm of natural gas annually from the USSR.

In 1982 Statoil made an interesting gas discovery at Tromsøflaket – the Askeladden field – but its location in northern Norway makes it commercially unviable. Given the distance from northern Norway to markets in Europe, substantially more gas will have to be discovered in the area – or elsewhere on the Norwegian continental shelf – to justify commercial development.

Several studies have been undertaken by Norwegian companies to assess alternative ways and means of transportation. These studies concern four main alternatives:

(a) A pipeline through Norway to the Continent.

(b) A trans-Scandinavian pipeline through Norway, Sweden and Denmark to the Continent.
(c) An offshore pipeline from Askeladden and along the coast of Norway to the UK or the Continent.
(d) A pipeline to land in northern Norway and a LNG terminal for the export of liquefied gas.

These studies have been completed and all the alternatives appear to be technically feasible, although more work will have to be done to clarify economic, organizational and other aspects.

The Sale of New Norwegian Gas. Deliveries of new Norwegian gas will have to come from fields such as Sleipner, Troll, Askeladden and others under development. Most of this new gas is in deep water and some of it in remote areas – for example the Troll field, located in areas of water between 300 and 400 metres deep. In the first phase of the development the partners envisage installing a single integrated platform around 450 metres high, with an annual production capacity of 15 mtoe of gas and 100,000 b/d of oil. Nothing like this has ever been constructed, and new technology will have to be developed. Investment in the platform, including production wells and sub-sea facilities, will be at least $5 billion (at the 1983 exchange rate). As mentioned earlier, the Askeladden field requires more gas to be discovered in the area before a pipeline is considered, but in any event it will be difficult and expensive to transport the gas. At the moment the sale of gas from the Sleipner field is being negotiated with Continental and British buyers. Transportation will be either by existing pipelines or by a new pipeline, with production starting in the first half of the 1990s. From a reservoir and development point of view this project is well defined.

An important question when discussing the future role of Norwegian gas is what the market would be willing to pay for it. A number of factors will determine this, the most important being:

(a) Total demand, including the division between residential and industrial sectors.
(b) The infrastructure of the market (the gas grid).

(c) Present gas supplies and availability of future supplies, as well as their diversification.
(d) Competition from other energy sources.

Over time there have been important movements in the distribution of gas consumption across sectors. In the 1960s and early 1970s consumption was fairly evenly distributed among the residential/commercial sector, the industrial sector and the power generation sector. A shift has gradually taken place towards the residential/commercial sector with a corresponding decline in the share of power generation. This move towards premium uses is important for the market's ability to absorb new high-cost gas.

During the 1960s and 1970s investments were made by the European gas distribution companies to develop an infrastructure and a gas market. Low-priced gas made it possible to penetrate the energy market, and now that natural gas is an established source of energy, the investments in the infrastructure have to a large extent been amortised.

Over the last few years the Norwegian reserve base of oil as well as gas has increased substantially. As far as oil is concerned, it is now evident that the oil in the Troll field is producible. Furthermore there are substantial oil reserves in the Gullfaks and Oseberg areas, and the discovery at Haltenbanken is also significant in this respect.

However, regarding new reserves and deliveries from the Norwegian continental shelf, the gas will come from areas of deep water and/or with long distances to the market. It is therefore evident that new Norwegian gas will be high-cost gas, and will be marketed only to the extent that the buyers are willing to pay sufficiently high prices to justify long-term development and transportation costs, as well as a reasonable profit to the sellers. In this context the alternative of development of further oilfields will always be an important Norwegian option.

Conclusion

Over the coming decades there will be a steady increase in demand for natural gas in Western Europe. From the beginning of the 1990s there is likely to be an increasing market for new Norwegian gas in Western Europe, potentially continuing beyond the year 2000. In these circumstances deliveries of new Norwegian gas, both to the Continent and to the UK, should be possible and the Troll field, for example, could become an important element in the future development of the West European market for natural gas. However, the penetration of gas into the market will be dependent upon the price of gas in relation to those of alternative energies such as oil, nuclear power and coal. It must be emphasised that Norwegian gas will never be a cheap gas, partly because of high costs and partly because deliveries are based on long-term agreements from a stable area.

5 THE PROSPECTS FOR GAS IN DEVELOPING COUNTRIES*

Aman R. Khan

Introduction

The continuing increase in oil prices during the past decade and the recent world-wide recession have led to a large increase in the current account deficits of the oil-importing developing countries (OIDCs). The economic slow-down in the industrialized countries has also reduced the expansion of raw material exports from these developing countries, which are vital to pay for oil imports. The impact of higher energy prices on the OIDCs in the short term is a reduction in economic growth and increased borrowing of funds to finance oil imports. In the longer term, they are obliged to develop alternative energy sources as a substitute for oil, or to encourage exploration for domestic hydrocarbon supplies.

In the past, investment in hydrocarbon exploration and exploitation in the OIDCs was solely at the behest of the majors. Since 1973, this situation has drastically altered with the growing influence of OPEC and the decline of the international oil companies as a governing force in decision-making in energy supply. The decade of the 1980s, therefore, requires national energy policies which have to be carefully planned and implemented, not only for the industrialized countries but even more so for the developing countries. These energy policies are critical for continued economic growth, and the first step is an identification and assessment of national energy resources including domestic oil, gas, coal, shale, tar sands, uranium, geothermal, and hydropower. On a lesser scale renewable resources such as solar and biomass also have to be taken into account.

* Paper originally presented at the Oxford Energy Seminar, 1982. Data updated 1986.

The development of indigenous energy resources and alternative strategies designed to reduce oil imports requires integrated energy planning and appropriate institutional structures. Furthermore, such programmes require large capital investments which are scarce. According to the World Bank about $40 billion per annum will be needed in the next five years for energy programmes in the OIDCs, including $5 billion per annum for oil and gas. Although such investments are enormous by any standard, this should be compared with a net fuel import bill of $74 billion for the OIDCs in 1980. On the other hand, it is widely recognised that developing countries have a large potential for oil and gas production, and natural gas is presently an underutilised resource. Domestic natural gas has also a distinct advantage, since it remains one of the cheapest energy options in developing countries as a replacement for fuel oil. The objective of this paper is to analyse the role that natural gas can play in developing countries in alleviating their energy problems.

World Natural Gas Reserves and Production

The estimated commercial production of natural gas during 1984 increased by about 9 per cent over the previous year. Proven reserves world-wide showed an increase of 30 per cent in 1984 over the 1980 estimate. In comparison, proven oil reserves showed an increase of only 1 per cent for the same period, thus indicating the growing interest in finding more gas to replace expensive oil. Table 5.1 presents a breakdown of estimated reserves and commercial production of natural gas world-wide by major regions. The world reserve life index for natural gas is about fifty-six years as compared with approximately thirty-four years for oil.

The USSR accounts for 39 per cent of total world gas reserves and the Middle East about 26 per cent. The increase in proven reserves of natural gas in the world has not abated. In fact, gas reserves have doubled twice in the past two decades and presently amount to nearly 90 per cent of oil reserves in terms of energy equivalence. Production and consumption in North America have levelled off, but in Western Europe and Japan the demand for gas continues to grow and marketed

Table 5.1: World Natural Gas Reserves and Commercial Production.

Region	*Reserves 1st January 1985 (tcf)*	*Commercial Production 1984 (bcf)*
North America	294	19,976
Central and South America	192	2,736
Middle East	904	1,637
Africa	208	1,804
Western Europe	194	6,540
Far East	256	3,785
USSR/Eastern Europe	1,348	23,073
World Total	3,396	59,549

Source: *Petroleum Economist*, Vol. LII, No. 8, August 1985.

production of gas in the world has shown an increase of 50 per cent in the past decade.

Although the major production and consumption of natural gas is in the developed countries, gas has been discovered in developing countries and with the increase in prices of imported energy, efforts are under way to exploit the domestic resources of gas. The World Bank categorizes developing countries on the basis of their income and net trade in oil. Thus, even net oil-exporting countries with low or middle income can be classified as developing countries. Based on this categorization, a list of developing countries with natural gas resources is presented in Table 5.2. It should be noted that four of the countries listed in Table 5.2 are OPEC members but the vast majority of the developing countries depend in varying degrees on imported oil.

Natural Gas in Developing Countries

The developing countries presented in Table 5.2 include those which are self-sufficient in energy. Some are net exporters of oil at the present stage but, because of declining oil reserves, may become net importers of oil in the near future. Nevertheless, in all cases energy consumption is growing and the rise in oil

Table 5.2: Classification of Developing Countries with Natural Gas Resources by Region, Income Group and Net Trade in Oil.

Region	*Net Oil-exporting Countries*		*Net Oil-importing Countries*	
	Low Income	*Middle Income*	*Low Income*	*Middle Income*
Africa	Angola	Congo Nigeria	Sudan Tanzania	Cameroon Ivory Coast Ghana
Asia and Pacific	Burma Indonesia	Malaysia	Afghanistan Bangladesh India Pakistan	Thailand
Europe, Middle East and North Africa		Algeria Egypt Oman Syria Tunisia		Morocco Romania Turkey Yugoslavia
Latin America and Caribbean		Bolivia Ecuador Mexico Peru Trinidad and Tobago		Argentina Barbados Brazil Chile Colombia

Source: *Energy in the Developing Countries*, World Bank, August 1980.

prices has placed considerable strains on their economies. In the case of OIDCs this has been particularly true and these countries are finding it increasingly difficult to finance oil imports while at the same time trying to maintain economic progress. With the assistance of international agencies, many of these countries are looking at the development of alternative energy resources. One problem is the lack of information on the resource base, and the large majority have no systematic plans for the development and exploitation of the potential resource.

The proven gas reserves and present production rates for some of the developing countries are shown in Table 5.3. About

Table 5.3: Gas Reserves and Production in Selected Developing Countries.

Region	*Reserves 1st January 1985 (tcf)*	*Production 1984 (bcf)*	*R/P (years)*
Africa			
Algeria	109.0	1,360.6	80
Angola	1.9	12.0	158
Congo	2.5	0.7	—
Egypt	8.3	143.0	58
Morocco	0.1	2.8	36
Nigeria	47.0	97.1	484
Tunisia	3.0	15.2	197
Central and South America			
Argentina	23.8	493.0	48
Bolivia	4.6	92.0	50
Brazil	3.0	73.8	40
Chile	4.5	48.4	91
Colombia	4.6	143.4	32
Ecuador	4.1	2.0	—
Mexico	76.7	1,036.5	74
Peru	0.9	45.6	19
Trinidad	12.4	191.8	64
Far East			
Afghanistan	2.1	102.0	21
Bangladesh	13.0	148.3	88
Burma	10.6	22.6	468
India	16.9	113.7	148
Indonesia	38.9	1,061.2	37
Malaysia	49.1	324.8	151
Pakistan	21.2	354.5	60
Thailand	7.4	84.8	87
Middle East			
Oman	6.4	47.3	135
Syria	3.7	5.3	—
Turkey	1.2	—	—
Total	476.9	6,022.4	79.2

Source: *Petroleum Economist*, Vol. LII, No. 8, August 1985.

14 per cent of the world's gas reserves are located in these countries. Excluding the net oil-exporting countries, the balance of 98 tcf of gas reserves (3 per cent of the world total) is to be found in countries that are net oil importers. Those in the latter category are in urgent need of broadening their energy resource base, faced with growing oil import bills and balance-of-payments deficits. Some of these countries have established the beginnings of a gas industry, some are just embarking and some have gas resources that are presently undeveloped. The relatively low level of natural gas use in the OIDCs is a result of the lack of interest by the major oil companies in developing an energy resource which is not easily exportable. Also, in the past, the domestic market in these countries has been too small to justify the large investments required for development of this energy resource. However, with the rising cost of energy, particularly oil, and with the increase in energy demand, it is only appropriate that the international agencies assist these developing countries to devote more attention to the potential for development of their gas resources.

Development of Gas in the Oil-importing Developing Countries. The development of a gas industry can typically progress through the following stages:

(a) Exploration and discovery.
(b) Field development and establishment of reserves.
(c) Surface facilities – gathering systems and gas processing.
(d) Gas transmission to major load centres.
(e) Gas utilisation by industrial consumers.
(f) Gas distribution systems for domestic and commercial customers.

The initial stage of exploration activity requires a large investment and considerable expertise. The existence of a known sedimentary basin may or may not result in the possibility of finding oil or gas. The capital required for the initial stage may be beyond the financial means of the OIDC, and foreign private capital would require a negotiated agreement satisfactory to both parties. Private oil companies are reluctant to participate in such exploration efforts without the necessary incentives.

Once gas reserves have been established, the initial market for the gas will normally be in its use for power generation or large-scale industries where gas can easily displace more expensive oil imports. Identification and location of these major markets then determine the capacity and size of surface facilities and transmission lines supplying the gas from the field to the load centre. When natural gas supply to major industrial consumers has become a reality, consideration can then be given to gas distribution systems in towns and cities to meet domestic and commercial loads.

The above scenario for gas development in an OIDC is logical and rational since there are large investments required for production, transmission and distribution of gas, and large enough volumes of gas demand must exist in order to compete with petroleum products in the market-place.

Gas Utilisation. Assuming that sufficient gas reserves are proven, the replacement of imported oil or oil products by domestic supplies of gas will be principally as a fuel, or in processing where value-added products are obtained. Fuel replacement for large industrial loads such as electric power generation is the primary usage in the early years of gas development. In later stages as higher-value uses for gas are developed in the industrial or domestic sectors, power plants fuelled by natural gas can be converted to other fuels. As a feedstock in chemical processing, natural gas has diverse uses. Natural gas liquids can be removed from wet gas and the natural gasoline, propane and butane fractions sold separately. The ethane fraction can be separated and processed for ethylene and subsequently polyethylene production.

Natural gas is also a feedstock for production of fertilizer (ammonia and urea) as well as methanol. As a fuel it is used in the manufacture of cement, glass, bricks, ceramic, etc., and in the mineral industries its utilisation ranges from direct iron ore reduction, smelting and beneficiation to the production of aluminium by providing cheap electric power. A tabulation of selected industrial uses is shown in Table 5.4 and the gas requirements per unit of production are presented in Table 5.5.

Whereas all of the above uses are for stationary loads, the use of natural gas in the transportation sector is also a strong

Table 5.4: Possible Uses of Gas by Industries.

Industrial Sector	*Applications*
Electric power plants	Electrical prime-mover fuel or fuel for steam generation
Iron and steel	Iron and steel making
Metallurgy and metal fabrication	Electro and chemical metallurgy Foundries Non-ferrous and metal working Mechanical and non-mechanical manufacture
Chemicals	Feedstock and fuel for chemical plants
Glass and building materials	Cement Bricks, tiles and ceramics Glass making and working
Food	Food products and beverages
Paper	Wood processing, pulp and paper
Miscellaneous	Textiles and leather goods Extractive industries Treatment of non-metalliferous mining products other than coal Precision instruments, tools Steam generation Stationary prime-movers other than for electrical generation, etc.

possibility. Recently, Mobil has developed a process by which natural gas is converted via the methanol route to motor gasoline, and a plant is under construction in New Zealand for processing of Maui gas. The use of compressed natural gas (CNG) as a vehicular fuel in gasoline and diesel engines has been successfully proven in the USA, Italy and the Netherlands. A feasibility study sponsored by the World Bank is investigating the conversion of fleet vehicles from gasoline and diesel fuel to CNG in Bangladesh. It is estimated that there are between 400,000 and 500,000 vehicles world-wide burning CNG as a fuel. The cost of operation is about 5.0 US cents per mile and the payout on the conversion investment is estimated at about three years.

Finally, in the case of OIDCs with large proven reserves of natural gas, export of natural gas via pipeline or as LNG is

Table 5.5: Natural Gas Requirements for Typical Industrial Usage.

Industrial Sectors	*Gas Usage per Unit of Production*
Electric Power Plants	
Electricity generation	10–12 scf/kWh
Metal Industries	
Aluminium production	268.00 mcf/ton
Copper production	18.20 mcf/ton
Lead production	4.56 mcf/ton
Zinc production	26.80 mcf/ton
Nickel production	5.60 mcf/ton
Magnesium production	269.60 mcf/ton
Manganese production	5.26 mcf/ton
Sponge iron production	14.40 mcf/ton
Chemicals	
Ammonia production	34.50 mcf/ton
Methanol production	35.30 mcf/ton
Carbon black production	17.60 mcf/ton
Glass and Building Material	
Cement production	6.9–7.45 mcf/ton
Glass fabrication	10.0–15.0 mcf/ton
Brick, tile, ceramic fabrication	8–10 scf/1,000 pieces
Paper and Pulp	
Market pulp production	5 mcf/ton
Kraft paper and board production	5–12 mcf/ton
Newsprint	7–10 mcf/ton

a possibility. For exports of 300 mcf/d as pipeline gas or LNG over a period of twenty years, about 2.5 tcf of natural gas reserves would need to be dedicated to the project.

Financial Aspects

The development of a gas industry takes time and requires a long-term commitment and detailed planning. More important, in the case of a developing country, the capital investment required for such a project is very large and sometimes beyond the country's internal financial capability. Several alternative

Table 5.6: Selected Developing Countries: Comparative Costs of Natural Gas Production.

	$ per Barrel of Oil Equivalent	*Opportunity Cost of Alternative Fuel at Appraisal Time ($ per Barrel)*
Associated Gas (Offshore) Delivered to Power Station (Egypt)	6.0	Fuel oil 19.5
Non-associated Gas Offshore Delivered to Power Station (Thailand)*	12.3	Fuel oil 15.4
Non-associated Gas Onshore Delivered to Power Plant and Fertilizer Plant (Bangladesh)	4.0	Fuel oil 21.8
Non-associated Gas Delivered to Residential Consumers (Egypt)	17.0	LPG 48.0

Note: * Gas produced by an oil company; includes return on equity.

Source: World Bank staff estimates.

sources of funds are available: these include suppliers' credits, funding from the World Bank and other international lending agencies, commercial bank credit and export–import credits tied to purchase of equipment. The World Bank has taken a leading role in energy-related lending programmes in recent years and has provided funding for the development of natural gas in Thailand, Pakistan, Egypt, Latin America and Bangladesh. Some OPEC countries have provided funds, particularly where the projects have been demonstrated to be sound and the plans were well defined. Private investment is another source of funding, but such funding varies from country to country according to the incentive provided to private companies.

In a recent study, the World Bank compared gas production costs for selected developing countries with the cost of fuel oil replaced (see Table 5.6). The cost of gas production shown varies from $0.67/mBtu to $2.83/mBtu depending on the nature and location of the gas reserve. The major portion of the investment required in the development of a new gas field in

OIDCs is in the field development and drilling. The gas quality determines the investment required for processing, and the distance to the market dictates the investment required for the transmission line. A typical example of the capital investment required for the development of a new gas field in an OIDC in South Asia is presented below. The field concerned is located on shore, 50 miles from the market area and is assumed to produce 185 mcf/d of gas with a heating value of 900 Btu/scf:

Investment	*Million $*
Field Development and Drilling	131.5
Gathering Systems	2.5
Processing Plants	40.2
Transmission Line	20.5
Total	194.7
Gas Cost	$1.26/mcf
	$7.56/boe

The gas cost does not include royalties, taxes or the costs incurred prior to discovery. Nevertheless, non-associated gas onshore can be produced at 10–25 per cent of the price of imported oil. Obviously, offshore gas will be more expensive, with perhaps double the onshore production costs, because of the greater investments required for offshore platforms and submarine pipelines.

The price of natural gas on the world market until 1985 was less than that of crude oil or oil products on an energy-equivalent basis. This undervaluation of the price of gas in relation to other energy forms provided an inroad for gas into the world energy market. Crude oil at $34 per barrel is equivalent to almost $6 per mcf of gas at a heating value of 1,000 Btu per scf. However, natural gas prices in the world market have not yet reached this level, except in the case of LNG imported into Japan. Locally produced natural gas is priced far below the equivalent crude oil price, and thus natural gas has a distinct advantage over oil in the market-place. In the case of OIDCs where natural gas has been discovered, it is obvious that the utilisation of gas in place of imported oil will result in considerable savings of hard-earned foreign exchange. On the other hand, natural gas exported via pipeline or as LNG must compete in the market-place with crude oil. If the natural gas in the importing country is priced higher than crude oil on an energy-equivalent basis, it is certain that consumers will not accept the higher-priced gas. Therefore, there must be pricing parity of natural gas with crude oil in the market-place and not at the point of export.

At present, there is no unified pricing policy for exported gas. Pipeline export gas prices vary in different parts of the world as can be seen below:

From	*To*	*Border Price, $/mBtu*
Mexico	USA	No deliveries (1985)
Canada	USA	2.92 (1985)
Bolivia	Argentina	4.75 (1984)*

* Includes bartered goods and services.

The export project to supply Bolivian gas to Brazil contemplates a border price of $4.50/mBtu while the supply of gas from Algeria to Italy via pipeline has been stalled by the dispute on pricing, with the Algerians demanding a new f.o.b. price of $5.00/mBtu which is higher than the original contract price. Similarly, LNG prices f.o.b. vary for different export projects, since the negotiated price is usually calculated and indexed to a basket of crude oil prices, and transportation costs vary depending on the different distances in individual projects.

The price of gas for domestic usage in developing countries varies depending on the agreements with producers and the markets served. In Indonesia, the price of gas for petrochemical usage ranges from $0.50 to $1.00/mBtu while other industries are charged up to $2.00/mBtu. In Thailand, production from the Erawan field is sold by Union Oil to the Petroleum Authority of Thailand at a price of $2.20/mBtu and mainly used for power generation. In Latin America, the price of gas for internal usage is generally below $2.00/mBtu although large industrial customers (power plants) pay between $0.50 and $1.50/mBtu. The current gas price for industrial consumers in Bangladesh is between $1.50 and $1.71/mBtu

(1984). In Pakistan residential customers paid $1.20/mBtu while industrial consumers were charged $1.50/mBtu in 1984.

As a matter of policy, the delivered price of natural gas in OIDCs for domestic usage should not be higher than the price of the current fuel being delivered to the customer, nor higher than that of the next best alternative fuel. Also, it is axiomatic that the revenue flowing to both gas-producing and supplying companies or agencies should be high enough to cover their full accounting costs, including depreciation and sufficient return on capital to keep them financially viable.

Conclusions

In summary, the following points must be emphasised. Developing countries with net imports of oil must develop alternative energy sources whenever the domestic supply is likely to be cheaper than imported oil. In many cases, the most promising alternative is natural gas. The market potential for natural gas in the developing countries remains favourable for a variety of reasons related both to the prospects of higher oil prices in the 1990s and to the natural advantages of gas. These advantages include the known and increasing quantity of gas reserves and their recoverability, the ease of transportation, storage and the substitutability of gas for other fuels, as well as the current underpricing of gas in the market.

It is imperative that the current opportunities for the development of natural gas in the OIDCs be taken advantage of quickly, especially in the light of the long lead time and heavy investment required to bring a discovered field into production. A number of countries can reduce their dependence on oil imports significantly in the coming decade, and estimates indicate that production of natural gas in these developing countries could triple by 1990. However, this will require a considerable commitment on the part of the OIDCs and funding sources. The total investment required for the development of a gas supply on that scale could be in excess of $50 billion over the next decade, corresponding to an average development cost of about $13,000 per barrel of oil equivalent per day. Such high-cost investments may appear wasteful considering the recent fall in oil prices (1986) but, because of long lead-in times, they should be assessed in terms of the likely behaviour of petroleum prices in the mid-1990s. Future price rises would justify them fully.

Governments of OIDCs, for whom the development of oil substitutes is crucial, must therefore provide the economic incentive to attract private investment, both internal and external, while also investigating the availability of alternative sources of development capital. Careful planning must be employed to ensure the most economic use of the gas to be produced. Finally, the gas eventually produced must be allowed to compete on an energy-equivalent basis in the market in order to ensure reasonable pricing and the most economic utilisation of this presently undervalued resource.

6 THE EVOLUTION OF MARKET-RESPONSIVE PRICING FOR NATURAL GAS IN THE USA AND ITS EFFECT ON GAS TRADE*

James T. Jensen

In 1980, following the second oil shock, controversy erupted in international gas trade as governments attempted to assess how natural gas prices should respond to rising oil price levels. Producer governments argued that natural gas was a premium fuel and should command a premium over oil. They also argued that the high costs of transportation, either as LNG or by pipeline, should increasingly be passed on to the buyer to enable the producer to recover a more nearly equivalent energy value to oil at the well-head. Consumer governments contended that natural gas was not worth such a high premium in their own national markets.

The debate was largely over pricing formulae which would be used to establish precedents for price settlements between producing country and consuming country. The Algerians, who were perhaps the most aggressive in negotiating on behalf of the producer point-of-view, advanced the position that natural gas should be priced on a parity with their own premium quality crude oil f.o.b. Algeria. Such a formula would have priced regasified LNG in France, for example, at 122 per cent of the average cost of imported crude. Japan, which had traditionally placed a high premium value on LNG in its own markets, had accepted the concept of c.i.f. parity with crude oil in negotiations with Abu Dhabi. Most consuming governments resisted the producer claims.

The debates dragged on for an extended period of time. In January 1980, the Algerians requested a price increase from their French, Italian and US purchasers. By April of that year, the El Paso project to the USA had been suspended, ultimately

* Contribution specially written for the collection, 1986.

to be terminated. By February 1982, France finally negotiated a settlement with Algeria in which Gaz de France paid a price that it deemed to be commercially acceptable, while the French Foreign Ministry negotiated a 'political' premium in addition because of Algeria's position in the French community. The resultant settlement effectively priced regasified Algerian LNG at 95 per cent of average French imported crude costs. The French precedent was extended in modified form to the Algerian/Italian trade, but before it could be pushed as a more general formula for gas pricing, the growing weakness of world oil prices began to transfer the balance of negotiating power to the buyers' side.

The initial Algerian proposal was the 'high-water mark' of natural gas pricing formulae for international trade. Furthermore, the debates over whether natural gas should be priced at c.i.f. or f.o.b. parity *vis-à-vis* crude oil have all but ceased. In fact, in recent renegotiations and settlements between European buyers and the Soviet Union, Ruhrgas in West Germany was in the forefront of arguing that oil prices no longer dictated gas pricing, but that gas instead should be priced on the basis of coal competition. The dominant theme in international pricing discussions now has become market-oriented as distinct from formula-oriented.

Perhaps nowhere has this transformation in thinking from government-dictated to market-derived pricing been more apparent than in the USA. The turning-point in US policy came late in 1978 with the passage of the Natural Gas Policy Act of 1978 (NGPA), more than a year before Sonatrach made its pricing request of El Paso. In the USA, the period from 1954, when the Supreme Court extended price regulation to the well-head, to 1978, when Congress passed the NGPA, was marked by increasing government intervention in gas industry pricing. The NGPA reversed that trend and created a markedly different environment for establishing prices for natural gas.

It is now generally recognised that the market clearing price for natural gas in the south-west producing areas of the USA is about 55 per cent – not 122 per cent or even 100 per cent – of the average refiner acquisition cost of crude oil. While imports landed nearer the major consumption centres would enjoy a transportation advantage – a 'place differential' – it is now clear that competitive markets will not support the level of import prices proposed in the early 1980s.

In 1975, when the USA accounted for 46 per cent of total world natural gas consumption, the country was in the midst of a serious natural gas shortage with supply curtailments imposed on many industrial and power generation customers, and restrictions placed on the addition of new customers in many parts of the country. Imports in 1975 totalled 953 bcf, or 4 per cent of total supply. While the predominant import was by pipeline from Canada, plans were under way to expand imports from Mexico by pipeline, to increase the existing LNG imports from Algeria and to include supplies from Indonesia, Trinidad, and Nigeria among others. Moves were also under way to bring in expensive supplementary supply such as the large Alaskan Natural Gas Transportation System to bring North Slope gas to the lower forty-eight states and plans to build coal gasification facilities for synthetic gas.

Ten years later in 1985, all LNG imports into the USA had ceased and the 911 bcf that the USA took from Canada had become its only remaining import. In fact, as recently as 1983, US imports of Canadian gas were only 42 per cent of the quantity the Canadians had authorised for export.

The dramatic change can be attributed to two factors. The first of these was the sharply reduced expectation for growth in gas demand and the resulting lessened need for gas supply. This change is attributable to the higher energy prices and their effect on conservation, structural change in energy use, and fuel-switching.

The second major factor has been the emergence of increasing policy emphasis toward market-responsive pricing in the USA as exemplified by Order 436 of the Federal Energy Regulatory Commission issued in October 1985 which forces the US gas industry into a much more competitive and market-responsive mode. Market-responsive pricing has now also been adopted by Canada for implementation later in 1986 and the policy thrust in the UK and West Germany, among others, is now more market-oriented than it is price-formula oriented.

The Pricing Dilemma

One major characteristic which differentiates the market economies from the centrally planned economies is their reliance on the price mechanism for decentralized economic decision-making. Whereas the CPEs lay out central plans for production and commonly set prices administratively, the market economies for the most part prefer to allow the interaction of supply and demand to determine price levels in those cases where workable market competition exists.

In economic theory, the composite demand of all consumers taken as a whole will increase with falling prices while the composite supply on offer from competitive producers will decline. The price level where supply and demand achieve a balance is the market clearing price. If demand grows more rapidly than supply beyond that point, market prices will rise; if supply increases differentially, they will fall. Economic efficiency is a major benefit which flows from reliance on free markets.

These competitive market economics, however, apply to situations where workable competition is deemed to exist. Within the market economies, it is well recognised that natural monopolies occur in some sectors of economic activity, and hence the market mechanism is poorly suited to discipline prices. For these sectors, the governments generally abandon their free-market approach to pricing in favour of some form of administered pricing.

Electric utilities provide a classic example. The heavy capital costs and economies of scale in electricity distribution result in declining unit distribution costs with volume, thus effectively foreclosing competition from other distributors within a given region. Hence, electric utility service is everywhere treated as a natural monopoly provided either by government-owned corporations as in the UK and France, or by government-regulated private utilities as in the USA and Japan.

In sharp contrast, oil distribution does not impose similar barriers to competitive entry. Everywhere within the industrial countries of the OECD, multiple oil suppliers compete for markets, even where the government has itself established a national oil company.

In this pattern of selective monopoly regulation in an otherwise competitive market economy, natural gas has always stood out as an anomaly. Within the retail gas distribution systems, service to residential, small commercial, or premium industrial customers appears very much as a classic natural monopoly for which utility regulation is designed. The residential user, for example, has no alternative supplies of natural gas, and the costs of converting his heating system to oil or electricity are such that there is little protection against monopoly pricing.

But in the USA, as in some other industrialized countries, much of the natural gas is sold into large dual-fired boiler installations where the natural monopoly model does not really apply. Few things offer a greater deterrent to monopoly gas pricing than service to a dual-fired boiler customer, who can switch to an alternative fuel at a moment's notice if the price of gas is too high. Thus, natural gas, unlike electricity, does not comfortably fit the regulated monopoly model. And, unlike coal and oil, it does not fully fit the competitive commodity model. With a foot planted in each camp, natural gas policy-making in the USA has been a victim of what has become a clash of economic cultures.

The Trend Toward Increasing Intervention

Retail distribution in the USA, as in many OECD countries, originally began as sendout from local town gas plants. Gas distributors thus strongly resembled electric utilities and were initially regulated in much the same way. Local distribution companies have continued to be regulated by state public service commissions, even after their conversion to natural gas and their connection to the national grid system.

In US practice, rate regulation for utilities must be done on the basis of 'just and reasonable' principles. 'Just and reasonable' rates usually, though not always, provide an allowance for operating costs plus a reasonable return on historic investment. Rates so derived are cost-based and do not take account of the value of alternative fuels or the market implications of the particular pricing that is finally arrived at. This approach, known as cost-of-service rate-making, has been traditional in the utilities subject to state public utility commission regulation.

As the US inter-state natural gas pipelines began to develop, it seemed only logical to extend utility regulation to these systems as well. This was done in the Natural Gas Act of 1938 which placed the inter-state pipelines under the rate-making jurisdiction of the Federal Power Commission (then called the FPC but now called the Federal Energy Regulatory Commission or FERC).

Although the language of the Natural Gas Act appeared to exempt producers from price controls, the Supreme Court in its 1954 Phillips decision interpreted the law as requiring 'just and reasonable' rate regulation at the well-head if the goal of preventing monopoly pricing to consumers was to be achieved. Thus, well-head price regulation was officially instituted in the USA through another extension of the logic which first brought the inter-state pipelines under monopoly rate regulation.

It proved to be far more difficult to apply cost-of-service rate-making to a myriad of gas producers than it had to local distribution companies or inter-state pipelines. The FPC floundered with the problem for a time but gradually evolved an approach which first averaged costs on an area-wide, later on a national, basis. But the principle of cost-based regulation was retained. Well-head price regulation, however, was selective. Since it applied only to producers selling to inter-state natural gas transmission companies, production for local use in the large intra-state markets of Texas, Louisiana and Oklahoma was not subject to FPC well-head pricing.

Thus, not only was price-regulated natural gas competing for market share against unregulated coal and oil, but inter-state pipelines subject to price regulation were also forced to compete for supply with intra-state lines that were not similarly constrained. Despite the presumption that the pipelines were regulated natural monopolies, the market continued to operate to influence the levels of both demand and supply available to the inter-state suppliers.

From the time of the Supreme Court decision in 1954 until 1972, when demand for natural gas in the USA peaked, natural gas consumption grew at a rate of 6 per cent per annum and it carved out a substantial market share against unregulated coal and oil. By the early 1970s, it was increasingly apparent that the low regulated prices were not stimulating adequate development of new reserves for the inter-state market. The unregulated intra-state market, rapidly growing as intensive industrial development took off in the Sun Belt, began to obtain the lion's share of new reserves which were becoming available. The inter-state market developed shortages and was forced to curtail gas supply to industrial and power generation users. By the early 1970s, reserve additions were inadequate even to sustain growth in the intra-state market and prices in this uncontrolled market rose rapidly under conditions of shortage.

During this period, it was easy to get agreement that the system was not working. What was difficult was to get congressional agreement on what legislation was needed to remedy the problem.

Advocates of free markets argued strongly for natural gas price deregulation to let competition set prices and determine the share of the US energy market which gas could capture. Advocates of stronger consumer protection argued just as vehemently for more comprehensive regulation ('reregulation'). They contended that gas price regulation did not go far enough since it exempted intra-state gas production from price controls. Not until 1978 could Congress agree on a direction for its new legislation and, after much painstaking negotiation, pass the NGPA. The NGPA accepted the thesis that substantial excess demand had been created for natural gas through price regulation and that the long-term solution to the problem was to move in the direction of competitive market-responsive pricing. It thus represented a watershed in natural gas price regulation. However, the NGPA's basic premise was that it would not be safe to deregulate natural gas while substantial excess demand existed. Therefore, the NGPA provided for a transition period to 1985 during which time price controls would remain on nearly all gas supply, inter-state and intra-state alike.

The NGPA provided for a large number of individual price control categories. Broadly speaking, gas discovered before 1977 was to be 'forever regulated' and not allowed to seek a market price, while a small quantity of high-cost, deep supply was to be totally deregulated. The remainder of the gas categories – so-called 'new gas' for the most part – was to be deregulated in 1985 when it was assumed that the incentive prices for this new gas would have offset excess demand by increasing US gas supply.

An Uneasy Transition to Market Pricing

A working assumption of Congress in formulating the interim price regulations in the NGPA was that competitive oil prices would slowly rise to a level of $15 per barrel by 1985. The second oil shock quickly destroyed that assumption and with it any early hope that international oil prices would provide any market price discipline for deregulated natural gas.

It had long been assumed by those who favoured natural gas price regulation that the consumer would get the full benefit from well-head price controls. It gradually became apparent that in this partially competitive natural gas market, supposed consumer benefits could also be transferred to suppliers. In an environment of perceived shortage where pipelines were precluded from using price competition for the bulk of their new supply acquisitions, they became quite undisciplined in price behaviour where price flexibility still existed. They utilised increasingly onerous contract terms to compete for supply where prices were fixed. Cross-subsidisation of imports, other supplements and, following the passage of the NGPA, deregulated deep gas became common. Since clearing prices were above average costs of pipeline supply, new expensive gas could readily be averaged (or rolled in) with price-regulated supply and still not forfeit market share. At one point following the passage of the NGPA, some pipelines were purchasing deregulated deep gas in the Anadarko Basin of Oklahoma for $10/mcf to resell in a market which on average allowed for a price of no more than $3. This cross-subsidy behaviour became increasingly flagrant following the passage of the NGPA and has led to still greater pressures to deregulate the system and push it further in the direction of market-responsive pricing than even the NGPA envisioned.

Before deep gas was deregulated, the only price flexibility which the pipeline had was in competition for imported gas. Thus, the signals to early pipeline and LNG projects were not market signals but subsidised price signals created by the price control of domestic gas.

Natural gas shortages developed in early 1977 as President Jimmy Carter first took office and were deep enough to have an impact on the general public and thus on political policy-making. The shortages were one important element in Carter's decision to rush forward to Congress a comprehensive National Energy Program. The President's proposal for natural gas favoured reregulation rather than deregulation. While it alluded to the possibility that some day natural gas might be deregulated, it proposed capping the price of intra-state and inter-state gas alike at the equivalent price of domestic crude oil, then $1.75/mBtu. Since domestic crude oil was itself price-controlled at a level of about 69 per cent of the refiner acquisition cost of imported crude oil, the effect was to maintain domestic new gas prices well below parity with c.i.f. crude oil imports.

The programme ran into early trouble. While the House of Representatives quickly passed something closely resembling the President's version, the Senate took much more time in its deliberations and made fundamental changes to the proposals.

The natural gas portion of the legislation encountered major problems as the Senate debated which approach to take. The new gas cap price was at the equivalent of $1.75/mBtu in the President's proposal. Shortly after the announcement of the President's programme, Canada unilaterally raised its price for exports to the USA to $2.16; but even greater difficulties followed the announcement in early summer that six US pipelines had negotiated a 2 bcf/d gas import project from Mexico and had agreed to a price equivalent to distillate prices in New York Harbour. Since distillate prices were also influenced by US crude oil price regulation, the net effect was that the Mexican imports would be at a level of rough parity with imported crude prices. This amounted to about $2.50/mBtu, well above the proposed cap price for US production.

Energy Secretary James Schlesinger quickly found himself in the midst of political crossfire on Capitol Hill. On the one hand, he had to explain to regulation-prone Senators why the US Export/Import Bank should be financing a Mexican project whose pricing terms did not meet the 'just and reasonable' standards which the President's programme presumably applied to domestic gas. At the same time, he had to explain to deregulation-oriented Senators from producing states why a price freely arrived at between a consortium of six US pipelines and Pemex was acceptable for imported gas but would not be reasonable for US producers. Faced with this dilemma, the Administration directly intervened in the pipeline/Pemex negotiations and the Mexican Government subsequently withdrew the project. While the US Government had always had import approval authority through the FPC and later through the Economic Regulatory Administration (ERA), this marked the first time that the US Government had become directly involved in import negotiations.

Although the Mexican project later started up at a much reduced level, it has not been a major source of US supply and was finally suspended in 1984 as gas prices were falling. It does not appear likely that imports from Mexico will be resumed in the near future.

The Senate's version of energy legislation was quite different from that which had been passed by the House. It took a full year-and-a-half for natural gas compromise legislation to be agreed upon between the Senate and the House in a conference committee. The NGPA became law in November 1978. Over that intervening period, deregulation won out over reregulation and the NGPA reversed the earlier historic policy direction. Although the timetable of six years for phasing in new gas deregulation provided a long transition period, the focus had dramatically shifted, from one in which Government set prices administratively to one in which the market was ultimately to prevail.

'Regulated' Deregulation and Runaway Oil Prices, 1979–82

The second oil price shock closely followed the passage of the NGPA. As oil prices ran away from regulated gas prices, the theoretical level at which the market would clear rose as well. The inter-state pipelines, which had experienced shortage and curtailment throughout much of the 1970s, were influenced by a perception of continuing shortage. They aggressively moved into gas acquisition programmes to alleviate their seeming problems. Many pipelines behaved as if price was of little concern provided they got the gas supply they needed.

Unfortunately, the industry as a whole had been largely out of the market during the 1970s because of shortages and therefore failed to understand that conservation, industrial restructuring and fuel-switching had brought supply and demand back into rough balance. The companies' aggressive supply acquisition programmes were therefore for a market that had largely disappeared.

In the climate of aggressive supply acquisition, the prices the pipelines were prepared to pay for imports came under closer Government scrutiny. In early 1980, the ERA was in the process of approving the increase in the El Paso/Sonatrach pricing formula that the escalation clauses under the old contract would have permitted. The approved price would have taken regasified LNG to the level of 65 per cent of the imported refiner acquisition cost of crude oil. When Sonatrach made its new pricing request of El Paso in early 1980, the precedent of direct US involvement in import negotiations was already established and Government officials joined the discussions. By February 1981, when the USA and Algeria failed to agree on a new LNG price formula, talks were broken off and El Paso ultimately terminated the project.

The termination of the Trunkline LNG project from Algeria took place somewhat later under different circumstances. Trunkline's original contract and ERA import approval both pre-dated the second oil shock. But by the time the project started up in late 1982, market resistance among Trunkline's customers to the approved price was very strong. When Trunkline suspended the project in December 1982, it cited adverse market losses as its reason for termination.

Canada had developed its own formula for pricing exports to the USA based on parity with imported crude oil in eastern Canada. Canada had been applying this 'substitution value formula' unilaterally although its decisions were still subject to ERA approval. With the 30 per cent Canadian oil-driven price increase in early 1980, the US Government again intervened directly. The Duncan–Lalonde formula, named after the energy ministers of the two countries, set agreed bounds to the price increases. When the formula was again adjusted upwards to \$4.94/mBtu in early 1981, it became apparent that the US market would no longer support further price increases. The price was initially frozen, then gradually lowered under market pressures. Formula pricing was finally abandoned when Canada adopted market-responsive pricing for exports in mid-1984.

By 1982, the pipelines' competitive efforts to improve their supply positions in this transitionally price-regulated market had led to a seemingly illogical array of prices for natural gas. In mid-1982, the average well-head price of natural gas in the USA was $2.50 or 46 per cent of the refiner acquisition cost of crude oil in the USA. But the average price of 'forever regulated' old gas was at a level of 25 per cent of crude at the same time that high-cost Section 107 gas, which included the deep, deregulated supply, averaged 137 per cent of crude oil. Canadian and Mexican imports averaged 93 per cent of crude.

Most of the gas that the industry acquired remained under price controls. Where the industry could not use price to compete for supply, it used tightened contract terms. It became common for pipelines to set take-or-pay thresholds as high as 90 per cent of well-head deliverability, a figure much above earlier experience. It also became common to write deregulation pricing clauses which had the effect of increasing prices to very high levels once new gas deregulation took place in 1985.

By 1983 and 1984, it became apparent that the industry had locked itself into volumes and prices which were increasingly unrealistic in the market-place. The pipelines' take-or-pay obligations began to balloon. As oil prices came down, residual fuel increasingly threatened to capture portions of the dual-fired boiler fuel market from gas.

As the 1st January 1985 target date for new gas deregulation approached, the gas industry found itself in the peculiar position of rapidly escalating well-head prices at a time when oil prices were dropping and demand for gas was falling rapidly creating a sizeable 'gas bubble'. Demand in 1983, for example, was 23 per cent below that of the peak year, 1972.

The market distortions of industry behaviour became increasingly obvious to the public, Congress and the regulators. If the thrust of policy as announced in 1978 was market-responsive rather than price-administered, the thrust of policy implementation since 1984, particularly under a Reagan

Administration that favours market forces, has been to accelerate the process toward deregulation.

FERC Orders 380 and 436

Faced with a series of inflexible supply contracts which could track neither declining market volumes nor falling competitive fuel prices, pipelines sought to preserve their markets by whatever means they could. Their normal response to ballooning take-or-pay obligations would have been to stop buying gas until natural field decline worked off the take-or-pay balances. The fact that the entire industry stopped buying gas created a large block of shut-in gas discoveries which the producers were anxious to sell. They attempted to find some way around pipelines' monopsony buying positions.

At the same time, large industrial users were finding that the pipelines' escalating weighted average system costs of gas were making gas uncompetitive with fuel oil. Their threatened switching to oil brought forth a series of special marketing programmes (SMPs) on the part of the pipelines. These provided initially for selective discounting to customers who threatened to switch but were clearly discriminatory to the large body of traditional utility customers of the distribution companies.

Pressures to transport gas between producers who wanted outlets and industrial users who wanted to avoid high regulated prices mounted. Where pipelines acceded to these transportation requests, they found they were losing load which could otherwise be used to work off their take-or-pay balances. By 1984, the system had become increasingly unstable.

Clearly, many in the industry hoped that Congress would pass legislation dealing directly with the problem of contracts between producers and pipelines. But Congress, as ever reluctant to deal with natural gas legislation, did not do so. The approach of the Administration and the FERC has been to give the distribution companies and end-users greater freedom to abandon traditional pipeline suppliers in favour of cheaper competitive offerings at the consumption end of the system. By putting the pipelines' market volumes at risk, FERC has hoped to force pipelines to renegotiate greater market flexibility into their supply contracts or even to abrogate them thereby creating market responsiveness at the well-head.

One early move in this direction was FERC Order 380 which relieved the distribution companies of any responsibility to pay minimum bills for commodity gas not taken. Thus, a distributor could freely shop elsewhere for cheaper gas if a pipeline's price were too high and not be obligated to honour the minimum bill it had originally signed on that contract.

Order 380 spelled the end of the last LNG import scheme into the USA. Distrigas of Boston, which was importing small quantities of LNG from Algeria, was forced into receivership because of the unwillingness of its purchasers to honour minimum bills for gas not taken in accordance with Order 380. Order 380 also put downward pressure on Canadian pricing which, although no longer operating under the rigid terms of the original Duncan–Lalonde formula, was still slow to adjust downwards to the US market pricing conditions. In July 1984, Canada announced a market-responsive pricing system for exports to the USA and followed it up in November 1985 with a decision to deregulate its own industry, instituting market-responsive pricing by November 1986.

In the USA pipelines have traditionally contracted for gas at the well-head (or point of import) and then resold it to distributors or end-users with the commodity and the transportation to market combined. Separate transportation has been, until recently, relatively uncommon. Following the passage of the NGPA, certain types of contract transportation were permitted and increased significantly.

As the pipelines began to utilise contract carriage (transportation) selectively to solve their market problems through the special marketing programmes (SMPs), the charge of discriminatory pricing was levied against them. At about the same time the courts disallowed the SMPs, FERC was already working on its own version of open, non-discriminatory transportation. This proposal was incorporated into FERC Order 436 in October 1985.

Order 436 provided that if pipelines offered transportation, as distinct from resale services, they would have to be open to all on a non-discriminatory basis; they also had to permit their resale purchasers to phase out of their resale contracts in favour of transportation contracts over a five-year period.

By permitting distribution companies and large end-users to negotiate with producers at the well-head and arrange separate transportation, FERC hopes to make well-head pricing competitive and therefore market responsive. Thus, increasingly, the distressed gas supply which could not find an outlet under the earlier system is free to compete in the market and set well-head clearing prices through direct negotiation between buyer and seller. While most pipelines did not initially adopt the option of becoming non-discriminatory transporters under Order 436, by the early summer of 1986 it was increasingly obvious that open transportation would be largely put in place before the 1986/7 heating season, thus leading to increased competition at the well-head as an influence on the price mechanism.

It has been increasingly apparent that many of the price distortions arose out of the fact that old, 'forever regulated' gas was being used to cross-subsidise above market level purchases. Since nearly all pipelines have system gas costs which are now at or above market clearing levels, old, 'forever regulated' gas no longer provides any cross-subsidy potential for new gas acquisitions. However, some cross-subsidy remains for a portion of existing gas under contract.

The Administration has wanted to decontrol this old gas to eliminate the last of this price-distorting element. Congress, however, has been reluctant to do so. As a result, there is a proposal for 'Administrative deregulation' which would allow FERC to raise old gas price ceilings. A positive decision by FERC on this subject would clearly eliminate much of the cross-subsidy potential remaining in the US system, but would require testing in the courts to see if, in fact, it would meet the 'just and reasonable' test of existing legislation.

Clearing prices in the South-west now appear to be at a level equivalent to 55 per cent of refiner acquisition cost of crude oil. There is nothing sacred about 55 per cent. That percentage would be expected to rise if the high distribution and transmission margins, which were selectively increased for the industrial markets during the 1970s, were to be reduced. This reversal of 'rate tilts' against the oil-switchable market is slowly taking place. Its effect will be to raise the natural gas well-head clearing price in the South-west as a percentage of the price of crude oil.

Imports of natural gas take place at border crossing points or LNG receipt terminals which are usually at some distance from the South-west and thus would be entitled to a place differential or premium above well-head prices. But even on the US East Coast where the premium presumably would be the highest, the value of imported natural gas is still well below parity with crude oil in US markets.

Conclusion

In the mid-1970s the USA appeared to be an excellent market for natural gas imports. Its own proved reserves were not being replaced; its market was growing rapidly; and price appeared to be no major obstacle to imports, since the industry could cross-subsidise new supply out of price-controlled old supply. These conditions no longer apply. US demand for gas has fallen dramatically with conservation, industry restructuring and fuel-switching. The pattern of well-head price controls which made it possible to pay above market prices for imported supplies is being dismantled, both through the congressional passage of the Natural Gas Policy Act of 1978 which opened the industry to long-term deregulation, and through Administrative moves made by the present FERC to make the market more responsive.

In this environment, the USA will not pay more for imported supply than the market will bear; this market is heavily influenced by a large group of dual-fired boiler users to whom residual fuel oil is the most logical alternative. Thus, the formula pricing debates of the early 1980s which equated gas value to c.i.f. or f.o.b. parity with crude oil are no longer pertinent. The pricing decision is now being made in the market-place by users who will pay no more for gas as boiler fuel than they will pay for residual fuel oil.

7 INTERNATIONAL GAS TRADE AND RESPECT OF CONTRACTS*

Ahmed Messili

During the last two decades the world energy scene has been in a state of turmoil, and the natural gas market is no exception. The increase in commercial production and gas utilisation, the growing interest in recovering associated gas and the search for new gas reserves have substantially changed the international gas situation. Nevertheless, the very high investments needed to gather and transport gas, along with their low rate of return and inadequate revenues compared to those of oil, continue to be serious impediments to any significant development in the gas market world-wide. In other words, an expansion in gas trade can only be brought about if the consuming countries accept that imported gas will not be cheaper than other hydrocarbons.

If a long-term energy supply–demand equilibrium is to be achieved, efforts to adapt to all these changes are necessary, not only from the producers but also from the consumers. In addition, the evolution of one source of energy can only be considered in the context of the whole energy scene. In this paper we propose to examine the development of the natural gas trade, the price changes and policies that it has engendered during the past years, and the outlook for the future.

Developments in Gas Supplies

For a long time natural gas consumption was restricted to a few geographical areas. Until the end of the 1950s, gas demand was important only in the USA, the USSR and to a lesser extent Canada. In the USA the share of natural gas in national energy consumption has been significant for a long time. In 1960 gas already accounted for one-third of the country's energy

* Synthesis of two papers presented at the Oxford Energy Seminar, 1983 and 1984.

consumption. This can be explained by a combination of several factors:

(a) The availability inside the USA of large gas reserves.
(b) A system of price regulation which encouraged consumption.
(c) Increased gas consumption by the industrial sector, particularly petrochemicals.

Faced with the prospect of depleted reserves since the early 1960s, the USA has increasingly had recourse to imports. At first these imports came from Canada and Mexico, but since then Algeria has become a supplier. However, these imports are not sufficient to prevent the depletion of reserves and maintain projected consumption levels. Therefore, either the Alaskan fields must be developed or LNG imports must be increased, probably from more distant sources. To satisfy increasing local demand the USA imported 25 bcm in 1981, and imports will probably reach 70 bcm by 1990, although the high level of investment required and the remote location of potential supply sources make it clear that an imported gas price equivalent to the domestic regulated gas price is not an attractive proposition to exporters.

In Europe the availability of domestic natural gas in large quantities is a recent development which has partly satisfied potential demand. Dutch (Groningen) and French (Lacq) producers stimulated a rapid transfer of gas to new users in Western Europe, but, faced with the rapid development of consumption and a supply which was soon found to be limited, gas companies negotiated long-term contracts. Imports originally came from geographically close sources (Algeria and the western part of the USSR), but since 1975 Libya has become a supplier and it is expected that Nigeria, Cameroon, Siberia and the Middle East will be involved in the future. Gas imports from outside Europe increased from 2.5 bcm in 1970 to about 34 bcm in 1981, of which 20 per cent is LNG. By 1990 imports are expected to reach 100 bcm, of which 38 per cent will be LNG, but problems will have to be faced here too. The development of gas in some of these areas requires large investments because of difficult offshore operations (for example in the Troll and north Norwegian fields), because of transportation over long distances through difficult terrain (Siberia) or submarine areas (intercontinental), or because the necessary liquefaction facilities are extremely expensive (it is estimated that the producer has to invest $1.2–1.5 billion for an LNG chain of 1 bcm per annum).

Japan, on the other hand, is a country depending almost exclusively on imports to meet its energy requirements, and since 1969 there has been a recourse to gas purchases in LNG form. Diversifying its supply sources to a maximum, Japan imports LNG from Alaska, Brunei, Indonesia and the Middle East, is currently negotiating supply contracts with Malaysia, Australia, Thailand, Canada and the USSR, and has succeeded in maintaining good relations with its partners by accepting an alignment of the gas price with that of crude oil. Its imports increased from 1.3 bcm in 1970 to 23.7 bcm in 1981, and may reach 57 bcm in 1990.

To conclude, the major consuming areas of the world will together require imports of about 230 bcm of gas by 1990. The total volume of gas trade could be higher if we take into account exports from Norway and the Netherlands to Western Europe, and those of Iran and Afghanistan to the USSR. More than 320 bcm would then be put on the international market in 1990 and around 400 bcm by the year 2000.

The High Level of Investments Required

The forecast expansion of the natural gas market will not take place unless large investments in field developments, liquefaction, transport and regasification are expended. In order to honour the contracts already signed or under negotiation, liquefaction facilities for more than 70 bcm of gas will be necessary by 1990. The cost of this equipment, including gathering systems, gas treatment units, compression stations and gas pipelines to the liquefaction plant, is estimated at more than $75 billion (1982 figure). In fact, from past experience, costs may be underestimated, for there have always been considerable discrepancies between the initial estimates and the actual final costs.

These cost levels impose a huge financial burden on those countries which dispose of natural gas reserves, and one which is often beyond their means. In the Third World especially, the seller does not get any profit from these very heavy investments, which require large loans to be raised from international financial institutions, including those of the buyer countries. Spelt out clearly, exporting countries are expected to develop the gas fields, lay gas lines (often over long distances), set up sophisticated liquefaction plants for which the technology is not yet entirely developed (and it should be remembered that this process consumes some 20–25 per cent of the processed gas), build expensive port facilities, and often participate in the shipping of gas by having their own fleets of LNG tankers.

Repayments of the contracted debt must of course be made on time, whatever the situation of the sale and purchase contract, and even in the event of a shut-down, for whatever reason. It must be added that anticipated construction costs are underestimated at the time the contract is signed, as are the terms of execution. So, the minimum price stated for most LNG contracts and assigned to cover execution costs is already exceeded when deliveries start, especially as costly foreign technical assistance is added to the operating costs.

It is true that the buyer also faces large investments, mainly for regasification plants and sometimes for LNG tankers, but the size of these investments does not compare with those of the seller. In the first place, it is the buyer country's industry that receives the greater benefit from these investments by providing the equipment and taking charge of their assembly, thus implementing advanced technology in a relatively new field. The seller's industry, on the other hand, receives a very low share in these investments. A European gas company official recently stressed all the advantages that issue from exports of technologies related to gas and LNG in particular.

In addition, for the selling country, the revenues expected from the exports often represent a significant share of its trade – when a developing country is involved – and are used to finance major social and economic projects. For the buying country, LNG imports represent a small part of its energy imports and a tiny share of its foreign trade.

It is obvious that these imbalances only increase the risks of destabilization and cause tensions detrimental to the implementation of a healthy contractual relationship.

Although the seller is protected in the case of failure by an extremely binding 'take or pay' clause, intended to make the buyer pay at the end of each contractual year for gas not lifted, this clause is difficult to enforce. In this event, the significant amounts involved become the subject of dispute between the two parties and it is generally necessary to submit the case to international arbitration, the proceedings of which are known, however, to be slow. Apart from the difficulty of enforcement, this clause is well-intentioned since it is the only guaranty for the seller. It is of the utmost importance that this clause is rigorously applied in practice, for otherwise it will be seen by producers as easily evaded.

An incident which may be a turning-point in the international gas trade, especially for LNG, is the decision (in December 1983) by a large importing company to suspend its gas liftings for an indeterminate period because of temporary difficulties. Another state-owned gas company has repudiated its contractual obligations, pleading the market situation. In both cases the terms of the contracts state without any ambiguity the nature of the obligations on each party and their imprescriptible nature. Clearly, anything less would have been unacceptable in respect of agreements for 20–25 years, reached after difficult and long negotiations and approved by the governments concerned. These actions therefore show the need to consider seriously the implications for the future development of natural gas and long-term world energy supplies. It is surprising that the international gas trade has not condemned these actions, which are detrimental not only to the seller, but also to the development of the gas industry in general, and LNG in particular. It remains to be seen whether any company will venture to invest further in such a cost-intensive and rigid industry, and whether any government will give its approval to such a costly commitment, incurring the risks of its plants being shut down, its facilities unutilised, and thousands of people standing idle. Clearly, this is an unacceptably high cost to pay merely because a partner is experiencing a difficult economic situation and alleging that consumption projections were incorrect at the time of the agreement, because of a shift in the importing country's energy policy, or because of the coming to power of a new government that questions past agreements. Media criticism of the gas policies of individual exporting countries cannot conceal the behaviour of some importing countries, nor reduce the apprehensions of potential LNG exporters.

Forecasts are unanimous in showing that gas requirements will be considerable in the future, but the development of natural gas hinges on a satisfactory solution to the problem of pricing. Decisions taken by some producing countries in the last few years to freeze or even cancel projects demonstrate the importance of establishing a pricing system that will allow the continuous development of the new resources, particularly in Africa and the Middle East, in order to compensate for the increasing deficit in supplies in the high-consumption areas which will begin in 1990. The determination to maintain the gas price at a low level, which has been expressed in some quarters, is in itself a real threat to the equilibrium of long-term gas supplies.

Evolution of the Price System

A Period of Low Prices. Until the beginning of the 1960s gas trade was limited to North America. In the USA the price of interstate gas was set for years according to complex formulae derived from the production cost and subject to the control of the Federal Power Commission. The fact that prices were kept at low levels contributed to its increased use, but soon led the producers to lose interest in the search for and development of new gas resources. Reserves reached their peak in 1967 and from then on imports from Canada and Mexico increased. Purchase prices for this gas evolved very slowly – the contract price for Mexican gas was set at 14 cents/mBtu in 1956 and has increased by only 1 cent every three years. During the 1960s the gas trade extended to geographical areas outside North America.

In Western Europe the discovery of the giant Groningen gas field enabled the Netherlands to negotiate important export agreements. The contracts they signed with France, Belgium, West Germany and Italy from 1966 to 1971 were based on prices ranging between 33 and 44 cents/mBtu. The Soviet export contract signed with Austria in the same period included a pricing system similar to that conceded by the Netherlands to its customers. In 1964, the first contract for LNG deliveries was concluded between Algeria and the UK, and provided for the sale of 1 bcm per annum over a fifteen-year period.

During the following years other contracts were signed for the delivery of Algerian LNG to Europe and the USA. In all of these contracts the gas pricing formulae referred to non-energy parameters, such as monthly salaries and the price index for steel. The selling of LNG was regarded as a banal industrial operation providing virtually no revenue for the producer, apart from the small revenue from taxation collected by the Algerian state. Such formulae allowed only a slow increase in prices (from $0.56 to $0.65/mBtu between 1965 and 1973) so that real prices were in decline as a result of inflation. This period of abundant energy supplies at constant and low prices and in close proximity to the consuming areas produced an average natural gas price higher than that for crude oil on an f.o.b. basis, but almost the same on a c.i.f. basis. It is worth noting that from 1970 some Western experts, foreseeing significant changes on the world energy scene, began to introduce formulae referring to the energy market into the new negotiations. Two years later this tendency spread among other gas exporters, and in 1973 the practice became general in response to the oil price adjustments.

A Period of Important Changes. The review of gas sales contracts, resulting from the oil price increase of 1973, affected countries as different as the Netherlands, Canada and the USSR. This new period for the gas industry saw the consolidation of trends observed in the previous decade: the doubling of the consumption rate on the one hand, and the development of new gas networks on the other. The escalation of the international gas trade, coupled with the demands of the exporting countries for price increases, led to a revision of the principles from which price formulae were derived. From this period contracts included a direct reference to the price of oil products, which linked the evolution of gas prices to the energy market.

New Criteria for Price Formulae. In the Netherlands the idea of reviewing gas export prices was widely supported by the Dutch economic press, which drew a parallel between prices charged to domestic clients (and for export) and the international price of low-sulphur fuel oil. The USSR also obtained gas prices indexed to fuel oil prices. Finally, from 1975 Canada set out to increase gradually its gas export price to reach the equivalent price of crude oil in 1977. In April 1977 Canada introduced the concept of substitution as the main criterion for gas price determination. This concept tallies with the cost of moving imported crude oil in eastern Canada. The Canadian objectives are clearly stated by the Secretary for Indian Affairs and the Development of the North in the following declaration made in 1975: 'The people of Canada must receive a reasonable return for their non-renewable resources while ensuring at the same time a sufficient return to industry so that exploration and development will continue.'[1]

From 1973 on, LNG exporters started to demand a reference to refined product prices in the contractual formulae. Algerian contracts were indexed to the prices of No. 2 and No. 6 fuels. Libyan contracts also referred to these prices as a complement to the earlier indices of building materials and salaries. In the case of Brunei, the price formulae included from the beginning a base price (for 65 per cent) and an indexation (for 35 per cent) on the price of a basket of crudes. In the contractual formulae agreed upon in 1976, 90 per cent of the price of the Indonesian LNG sold to Japan was determined by the price of a basket of nineteen Indonesian light crudes, and the remaining 10 per cent was to increase by 3 per cent per annum. Finally, the United Arab Emirates' LNG export contract price was set to follow the average price of Japanese crude imports on a calorific equivalence basis.

It can be seen, through these examples, that all the contracts in one way or another make reference to the energy market. An analysis of the formulae contained in LNG contracts referring to fuels enables us to draw the following conclusions:

1. The base price is implicitly considered as the c.i.f. price of a basket of fuels, less the transportation cost. However, owing

[1] *Wall Street Journal*, December 31st, 1975.

to the faster evolution of refined product prices with regard to transportation costs, the gap between refined product price and LNG price increased beyond reasonable limits. In a way, the logic of the formulae amounted to a 'transfer of rent' from the producer to the consumer.

2. The f.o.b. LNG price is indexed as a direct proportion of the c.i.f. price of a basket of fuels. However, when the price of energy doubled, the price of gas also doubled along with the gap. As a result, the difference between the gas price given by these particular formulae and the price of Arabian Light went up from 50 cents in 1975 to $1 in 1979, when the price of Arabian Light doubled.
3. The response to the changes in the adjustment mechanisms provided for in the contracts proved to be slow and ineffective. Combined with international inflation this resulted in a decline in gas prices in real terms.
4. The reference to refined products, the prices of which were defined within the framework of the consuming countries' energy and fiscal policies, amounted to granting the control of gas prices to the importing countries. In addition, the prices resulting from the formulae linking them to refined products were in turn subject to daily fluctuations that systematically disturbed the economics of the contracts, which were negotiated for long periods and called for heavy investments from both the producers and the consumers.
5. Finally, some important elements favourable to the producer do not appear in these contractual formulae:
   (i) The fact that the producer offers a guaranteed price and supply to the buyer is not quantitatively included in the pricing formulae.
   (ii) The exclusive reference to the least valuable petroleum products, such as low-sulphur and high-sulphur fuel oil, enabled gas to be used in the least 'noble' applications. The gas price policy was worked out in order to favour the industries of consuming countries, with the result that the producing countries were indirectly subsidising these industries.

The Increasing Inadequacy of the Formulae. As we have seen, the inadequacy of the various gas pricing formulae has been emphasised by the profound and rapid changes experienced in the oil market since mid-1979. During the first half of 1980, the gap between the LNG price and that of crude had become too large to be ignored. While the oil price had reached $6/mBtu, the Algerian LNG price fluctuated between $2 and $3.50. Natural gas and LNG exporters were naturally dissatisfied with this development and started to renegotiate their sales contracts in order to achieve more equitable gas prices.

A New Point of View on Prices. The alignment of the gas price with that of crude oil appeared to be the most appropriate means of avoiding continual fluctuations, offsetting the producers' investments, and stimulating the development of world gas trade. The opposition that this proposal attracted in Europe is well known, although in Japan the principle was readily accepted. This reference to crude oil is not arbitrary. It results from the change in the world energy situation and an awareness of the need to stabilize the value of gas hydrocarbons. Crude oil is, without doubt, the relevant energy price indicator for the following reasons:

(a) Its price is set by an organization that has shown a sense of responsibility in fixing the price of crude oil, taking account of not only the producers' interests but also the consumers' economic conditions.
(b) The crude oil price evolution is much more stable than that of refined products, and is less subject to speculation.
(c) Finally, the alignment of the gas price with that of oil is a necessary move for an energy source so long regarded not only as peripheral but also as possibly dangerous. The 'noble' application of gas brings it gradually closer to crude oil.

In fact, it was soon found that the resistance to such a development was more psychological than commercial, since the majority of the gas buyers now concede the principle of such a reference and no longer reject the principle of parity. Japan has seen the most rapid progress: all contracts with its LNG suppliers now refer to crude oil. From January 1980 the price of Abu Dhabi's LNG has been in strict parity with the price of Murban crude, on a calorific equivalence basis. A similar agreement exists for gas imported from Alaska, taking into account the average of Japanese crude imports. The price of Indonesian LNG to be delivered from the new Arun liquefaction lines is indexed 100 per cent to a basket of Indonesian light crudes. The price of LNG from Brunei is calculated according to the price of a basket of twenty crudes.

In the USA, there was an agreement in 1980 with Canada stipulating that the price of gas imported from Canada would follow the evolution of the Canadian crude oil imports price. In the same year a decisive change occurred in Europe when the Netherlands, under pressure from its Members of Parliament, adopted a new gas pricing policy in line with world trends. The USSR and Norway followed suit. In Algeria, the negotiations between Sonatrach and its LNG customers have resulted in agreements based on the principle of a gas price aligned with that of crude oil.

Nevertheless, importing countries continued trying to control the natural gas price. Taking advantage of stagnant energy consumption, some companies again tried to query the price formulae, asking for a considerable base price decrease and an end to the reference to crude oil and refined products on the international market. The modification of natural gas import regulation and its application to increasing contracts, along with the reference to strongly regulated and controlled internal gas prices, are all intended to dictate a price policy to the exporting countries. The imbalances in investments and incurred risks, and the freedom to dictate a price policy suited to the interests of the importing countries, would all result in a slow-down of natural gas development. Although not intended, this could also be the result of the EEC suggestion of a European LNG importers' coalition. Clearly, the desired supply security will not come as the result of a European gas-importing countries' cartel. On the contrary, the implementation of this plan in any form would inevitably lead to a deterioration in the relationship between producers and consumers.

The Future Role of Gas. Crude oil alone cannot meet world energy requirements despite the apparent abundance at present. Current forecasts of future energy consumption in the world, although carried out in a depressed economic situation and consequently underestimated, show that the role of natural gas, and particularly LNG, will not be negligible by the year 2000, and its share of consumption will remain at 18–20 per cent. This is natural since gas availability is increasing each year (reserves have been doubling steadily each decade since 1950), it is used in more and more diverse fields, oriented towards 'noble' uses (i.e. chemistry and transport) and, compared with other fuels, it is more flexible and has the great advantage of being a 'clean' fuel. The latter point is particularly important since industrial societies are increasingly concerned with environmental issues.

On the other hand, since gas facilities exist in large consuming areas, and it is easier and safer to develop than nuclear energy or even coal, we can assume that natural gas demand will be much greater than currently estimated. Projections of world energy consumption, socialist countries excluded, are generally based on an economic growth rate of approximately 3 per cent per annum and on a very substantial development of nuclear energy and coal. It is possible that the forecast economic growth rate is underestimated because of the current economic crisis, as were the rates forecast in 1978–9. Furthermore, development of nuclear energy and coal is probably overestimated for the following reasons:

(a) *Nuclear energy* – high costs and strong opposition in industrialized countries have already brought about a decline in comparison with programmes scheduled some years ago.
(b) *Coal* – severe pollution would result from an expansion of coal utilisation unless very expensive processing units (mainly desulphurization) were established.

It is possible that the relatively pessimistic nature of natural gas projections in relation to nuclear energy and coal are intended to hide the real fears of some consuming countries that they will experience a steady decrease of their reserves and production. In addition, it may serve to intimidate exporting countries threatening to switch to other sources. In any event, these estimates seem to indicate that the production/consumption deficit will continue to increase, mainly from the 1990s, and that the international gas trade will inevitably expand.

However, natural gas is not a competitor with other sources of energy, but rather a natural supplement to fossil fuels. Its characteristics and the range of its uses bring it closer and closer to low-sulphur crude oil, and are leading to the gradual replacement of crude oil by gas and increasingly numerous experiments on the use of gas for transportation. The development of its market should be seen in the framework of an optimal and rational exploitation of reserves, rather than as an attempt to create a cheap fuel to compete with heavy fuels or coal.

It is important to stress that the domestic production of large consuming areas (the USSR excluded) and that of their neighbours, which have provided them with natural gas, is going to decrease gradually. In order to get gas, these countries will have to go on to harsher and more remote areas which are more difficult to exploit. For example, it is estimated that the average distance to transport gas from producing areas to consuming areas could increase threefold in Western Europe by the end of the century, especially as consuming countries attempt to diversify their supply sources for security reasons.

By far the largest natural gas reserves in Africa are in Algeria, which is located at the crossroads of the large consuming areas. The first country in the world to have developed LNG, and to have laid an intercontinental submarine gas line (in partnership with Italy), Algeria intends to make its contribution to the world energy supply. In addition to European and US markets, the expected technical developments in LNG shipping would also enable Algeria to reach the Japanese market, which appears to be disposed towards a significant development of this energy source.

Conclusion

In conclusion, therefore, obvious potential exists for the development of the natural gas trade in the framework of a healthy and rational use of world energy sources. More than any other energy industry, natural gas, and LNG in particular, demands close international co-operation and involves long periods of time. The *sine qua non* for its success is a strict respect for commitments, agreed upon in a spirit of co-operation and recognition of mutual needs, and with no attempts at economic domination. Any approach based on short-term considerations or any attempts to set a low price for natural gas prejudicial to the producers' interests, will inevitably lead to the 'death' of a resource which is today comparable to crude oil by virtue of the size of its reserves.

8 PRESENT AND POSSIBLE FUTURE UTILISATION OF NATURAL GAS IN THE WESTERN PACIFIC REGION*

Ir Wijarso

The international trade of natural gas in the western Pacific region is relatively new compared with the development in other regions. Although it started in the late 1960s it only gained real momentum in the early 1970s. For the purpose of this paper, the region consists of Japan and Korea to the north, Taiwan, Hong Kong and the six ASEAN countries to the south, and Australia and New Zealand. Because of the geographical location of the countries in the region, the development of trade requires natural gas to be transported over large seas and oceans and the long distances separating the island countries prohibit the use of pipelines as a means of transport.

Only when large-scale commercially proven technologies for liquefaction and transportation of LNG became available in the late 1960s could there be trade of natural gas in the region. It is worth noting that in Indonesia the possibility of using methanol as a means of transporting the heat value of natural gas was also considered. Methanol has the advantage of cheaper transportation costs than LNG. Existing crude tankers can be used for transporting methanol, while the transportation of LNG requires new, expensive and sophisticated ships. However, this idea was abandoned because, in order to produce the methanol economically, units of about 5,000 tons would be required, and the plants then existing were not larger than 1,000 tons. Even today, the biggest methanol plant under construction is 2,500 tons. This plant will be used in the process of converting natural gas into gasoline in New Zealand.

Another factor that inhibited regional trade was the establishment of the sales price of natural gas. If the value of

* Paper presented at the Oxford Energy Seminar, 1984.

crude oil were to be used as the equivalent base, then, on an assumed average landed price of crude of \$1.80–2.00 per barrel, the natural gas price based on the heat value equivalent should then have been in the range of \$0.30–0.33/mBtu. This was definitely not sufficient to set up an elaborate liquefaction and transportation system and at the same time provide a large enough margin for an adequate return on investment. If, on the other hand, the price had been set higher it would have been too high compared with oil, and would probably not have attracted the potential users.

In the late 1960s, however, environmental pressures for clean air in Japan forced the power generating industry to look for a clean fuel. Low-sulphur fuel was one answer, which gave Indonesian Minas crude its competitive edge at that time, and large desulphurization plants were also being constructed. But natural gas was considered as a longer-term solution – some sort of transition fuel from oil into the nuclear era. For whatever reasons, this situation caused natural gas to be accepted as a clean fuel and gave it some sort of new leverage by which it could command premium prices. The first LNG shipment to Japan from Alaska in 1969 was delivered to Tokyo, a densely populated area with very stringent air pollution regulations. The international trade of natural gas in the region today consists mainly of LNG delivered to Japan as shown in Table 8.1. Japan will receive about 25.16 million tons or about 3.45 bcf/d of LNG in 1984. Countries outside the region contribute 3.02 million tons.

By comparison, the local use of natural gas, except in Australia, is relatively insignificant. Taiwan, Thailand, Malaysia, Indonesia and New Zealand together are utilising approximately 1.25 bcf/d gas domestically and Australia is utilising almost all of the gas it produces – about 1.17 bcf/d. This total of 2.42 bcf/d shows that Japan is utilising more than 58 per cent of all the consumed gas produced in the region. Future development plans may, however, change this composition and gas will increasingly be consumed locally. This differs from the pattern of growth in the other gas-consuming regions like the USA and Europe, where utilisation starts with the local consumption. The low level of local consumption, therefore, is indicative of the development stage of most of the producing

Table 8.1: Imports of LNG to Japan, by Country. 1984.

	Contractual Volumes 1984		*Realization Jan–July 1984*
	Million tons	*%*	*Million tons*
Indonesia	14.00	55.7	7.85
Brunei	5.14	20.4	3.22
Sarawak	3.00	11.9	1.70
	22.14	88.0	12.77
UAE	2.06	8.2	1.13
Alaska	0.96	3.8	0.56
	3.02	12.0	1.69
Total	25.16	100.0	14.46

countries in the region. The exception to this pattern may be Taiwan, where a peak production of 250 mcf/d was reached around 1977 and a declining natural gas production will be met by importing LNG in the near future. Natural gas today is supplying about 9.5 per cent of the region's primary energy requirements.

An important factor in the development of the international trade of LNG was obviously the price. If we consider the price of the first LNG shipment from Alaska in 1969, it shows that the oil equivalent was about $3.00 per barrel (c.i.f.). When compared to the then existing crude equivalent of about $2.00 per barrel, we find that the buyers were willing to accept a high premium of about 50 per cent. In 1973, the first Indonesian LNG sales contract still showed a c.i.f. price about 26 per cent higher than crude, although this was divided into an LNG element, which was 97 per cent of the crude f.o.b. price of Indonesian Minas crude, and the actual transportation cost. All the later contracts are at about the same parity, depending on the negotiations. While in principle the parties involved in a negotiation agree on crude oil parity, the question is where to place the parity: on an f.o.b. basis at loading port or at the burner-tip c.i.f. Agreement is usually reached between these two extremes. However, a more important question is whether these levels of LNG prices can still be the guiding principle for future contracts.

According to the Japanese MITI's long-term energy forecast, there may not be any future contracts for sales to Japan, at least until after 1995. MITI is limiting LNG imports for 1990 to 36.5 million tons, which is approximately the volume of all outstanding commitments. The new deliveries are forecast to be from Australia and Canada, while for 1995 the expected imports of 40.0 million tons will include the expected additional deliveries from the Soviet LNG project. This will keep the natural gas share at about 12 per cent of Japanese energy consumption, compared with the current share of 7 per cent. If the Canadian project fails, and if the Soviet project is further delayed, the potential LNG market for 1990–95 will be about 6.4 million tons. The only other possible markets in the region are Korea and Taiwan. Commencing in 1986, Korea will receive about 2 million tons of LNG from Indonesia, and future expansion in Korea could double this amount. As for Taiwan, the current projection in their long-term energy demand forecast calls for an import of LNG by 1991 of about 1.5 million tons per year. Under these conditions, it seems that changes in the price structure will not cause new demand to increase substantially, unless of course MITI is willing to deviate substantially from its present guidelines.

Possible sales outside the region will be to the US West Coast market. Indonesia has an eleven-year-old contract for sales of some 4.3 million tons of LNG to Los Angeles, but although the contract is still in force it has not yet been implemented. The difficulties are not confined to environmental problems and pricing because of the long distances involved, but are also due to the fact that the USA has abundant supplies of gas, both produced internally and imported from Canada and Mexico. It does not appear likely, therefore, that additional natural gas will be required from outside the Western Hemisphere before the mid-1990s, as present forecasts indicate that sufficient gas will be available until that time.

This indicates that in future natural gas should be considered for domestic utilisation rather than for export which would, in addition, enable oil to be released for export. Since oil is a globally traded commodity, this would be a feasible policy, and natural gas is therefore likely to remain a regionally traded commodity. Transporting LNG over long distances in tankers to other regions is likely to create price imbalances. On the other hand, the possibility of overland transport by pipelines in other regions, such as the USA and Europe, creates more favourable conditions for pricing. Another policy option, which should be considered by countries with abundant natural gas reserves and exportable coal deposits, is the possibility of exporting the coal and replacing it domestically with gas, even though coal may have a lower export value than natural gas.

What then are the present plans for domestic natural gas utilisation in the region?

Malaysia, with an estimated reserve of 40 tcf, presently plans to utilise some 1 bcf/d of gas in a three-phase programme in peninsular Malaysia, utilising the natural gas from offshore Trengganu. This plan entails the construction of a pipeline system from the Kerteh area on the east coast to Johor Bahru and Singapore in the South, to Port Kelang and Port Dickson in the West, and to Prai in the North. Included in this natural gas utilisation plan is a 900 MW power plant and a 600,000 ton steel plant. In the Sabah area, the plans call for the utilisation of some 90 mcf/d for a methanol plant and a direct reduction sponge iron plant. The Sarawak gas reserves are mainly employed for liquefaction and, to some extent, by the ASEAN fertilizer plant located in that area.

Thailand, currently producing 220 mcf/d of natural gas, expects to double production next year to some 440 mcf/d, and by the end of the decade to step up production to 1 bcf/d. Its basic policy is to convert the natural gas into LPG, CNG, condensate and tail gas to replace oil which it would otherwise have to import.

Australia is planning to increase its natural gas consumption to some 1.84 bcf/d in the early 1990s to replace its fuel oil use, at the same time exporting 6 million tons of LNG.

New Zealand's plans are such that it may double its current use of natural gas by the early 1990s.

Indonesia, with over 80 tcf of natural gas reserves, is gradually increasing its domestic utilisation. An ambitious plan is now being considered to bring part of the 40 tcf of Natuma natural gas (located in the South China Sea), through Batam Island to central Sumatra, some 1,200 km away, to be used in the Batam industrial project and to fuel the steam generators for an enhanced oil recovery project. This first phase will involve some 500 mcf/d, but this is not expected to be in place before 1991. A further development may be the continuation of the pipeline through south Sumatra to Java because of supply constraints. Java is today using only some 200 mcf/d of natural gas. Opening up the possibility of using natural gas in the 1990s will certainly increase its use in Java to at least 1 bcf/d. Other smaller local fields are increasingly used to meet the local energy demand; for instance in north and south Sumatra, east Java, east Kalimantan and soon also in Sulawesi. By the end of the current five year plan in 1989, these local uses will increase the current consumption of 690 mcf/d to some 880 mcf/d of natural gas as energy.

By the mid-1990s, the regional consumption of natural gas may reach some 12.15 bcf/d, of which 1.30 bcf/d or 11 per cent will be imported from outside the region – a substantial increase from the present 7 per cent. However, the producing countries' own consumption will have increased by some 150 per cent.

In conclusion, I would like to make the following observations:

(a) The facts support the view that natural gas is basically a regional trade commodity. Conditions in other regions, however, may dictate different approaches, especially in the pricing policy.
(b) The regional trade in natural gas will reach saturation point within the next decade. As Japan is the only large importer of natural gas in the region, its Government's policy of restricting natural gas imports to 12 per cent of its energy mix plays an important role. No major increases in

sales can be expected unless this policy constraint is changed.

(c) The crude parity pricing principle for natural gas so far implemented in the regional trade seems to be the correct approach.

(d) The medium-term utilisation of natural gas, therefore, will basically be local consumption, because of limited export capabilities. The replacement of oil by natural gas and meeting future energy requirements with natural gas should be, as far as possible, the guiding principle for natural gas-producing countries. Oil still has a greater flexibility in international trade than natural gas. It may also be worth considering the use of natural gas instead of coal in countries with abundant natural gas reserves and exportable coal deposits.

In the past gas was considered to be a premium raw material, and therefore limited to 'noble' uses only. Perhaps its only noble use ultimately will be the money it can generate for the producing country.

9 GOVERNMENT INTERVENTION AS A FACTOR IN INTERNATIONAL GAS TRADE: HELP OR HINDRANCE?*

Jonathan P. Stern

Government intervention in domestic gas industries, and specifically in the domestic pricing of natural gas, is a familiar phenomenon. This paper will deal with the neglected, but increasingly important and visible subject of government intervention in international gas trade, and mainly with intervention by the governments of importing countries. A large number of exporting countries sell gas through national oil and gas companies so that the issue does not have the same relevance. In citing a number of different types, and different instances, of government intervention, the major argument of the paper is that, however it may be viewed by the commercial interests, this factor will need to be increasingly taken into account by importers and exporters in the planning of international gas trade.

Types of Government Intervention

(*a*) *Refusing to allow an importer to sign a contract which has been agreed between the commercial parties*. The most obvious and recent example of this in the UK context is the refusal of the Government to allow gas imports from the Norwegian Sleipner field to go ahead after a draft contract had been agreed between the British Gas Corporation and Statoil. The reason given was that the Government considered that sufficient additional gas had been identified on the UKCS to make imports from Norway unnecessary. In this instance, the Government took a different position from the utility on the quantity of available

* Paper presented at conference on 'The Planning and Financing of International Pipelines', Scientific Surveys Ltd., London, November 1985.

reserves, the time frame in which they could be developed and the price at which they could be brought to the consumer.

An example from Continental Europe would be the refusal of the West German Government to allow the import of gas from the Norwegian Ula field by Deutsche BP/Gelsenberg in 1981. The objection to this contract was the price, which was said to approximate to c.i.f. parity with crude oil, a principle which was unacceptable to the West German Government at that time.

An example from the USA would be a series of Government refusals, in the late 1970s, to give permission for base-load LNG imports from Algeria, on grounds of a lack of regional and national need for the gas. This was followed by a refusal to allow a base-load pipeline import from Mexico on grounds of price (see below under type (*c*)).

Another slightly different example, which brings in LNG trade, concerns the Algeria–Spain project. The terms of this contract were renegotiated at government level in 1985, after referring the case to the International Chamber of Commerce. Deliveries began in 1980 but have never risen above one-third of contractual volumes, mainly due to the failure of the Spanish Government to allocate sufficient funds to build the transmission system necessary to create additional markets for the gas. (From the side of the importer, the Algerian demands for higher prices have been given as a reason for failure to take a proportion of the volume.) Agreement between the Governments should allow full contractual volumes to commence from 1988.

(*b*) *Instructing a utility to sign a contract which it would not otherwise have considered commercial*. The prime example of this in pipeline gas trade has been the Italian Government instructing Snam to sign the contract for the Transmediterranean pipeline gas, despite the fact that the utility made it known that it considered the Algerian price to be unrealistic. In 1983, the utility forced the passage of legislation ensuring that the Government could be held to this commitment until the renegotiation of prices which will take place in early 1986. These events followed similar actions by Gaz de France during 1980–81, in its Algerian LNG contracts, where contracts were signed under pressure from the French Government, which agreed to contribute a subsidy equivalent to 13.5 per cent of the purchase price (a commitment it rapidly reneged upon).

The rationale for these actions by governments was primarily political in the French case and primarily economic in the Italian case. The history of Franco-Algerian relations created a climate in which the new Socialist Government of François Mitterrand felt that successful relations with Algeria could constitute the key to French relations with Africa, OPEC and the Third World in general. The Chadli Government made it clear that there could be no improvement in bilateral relations between the countries without a settlement of the gas contract to Algeria's satisfaction. In addition, the Algerian Government made it clear that agreement on a gas price and other contractual conditions was essential if French companies were to receive a large share of future Algerian industrial contracts. This latter point was the major reason for the Italian action, with the additional spur that Italian companies were owed $1.5 billion for work that had already been completed and did not look as if they would receive any payment while the issue was outstanding. It is useful here to distinguish between bilateral political as opposed to bilateral economic relations, even though the two may be importantly linked.

(*c*) *Taking over the commercial negotiations*. The only examples of this have been in the USA, where the Department of Energy was called in to negotiate directly with Algeria in the case of the stalled El Paso 1 LNG project and also in the import of Mexican pipeline gas. (While France, Italy and Spain have had high-level political discussions with Algeria over gas prices, the utilities have been responsible for negotiations, even if they were unwilling to accept the result.) The main reason for taking over the negotiations has been regulatory complexity and confusion. In the Algerian case, the reason for direct government intervention was primarily that the importing company could not negotiate meaningfully unless it could be sure that the result would be acceptable to the regulatory body. In the event, the negotiations failed and the project was shelved.

In the case of Mexican pipeline exports, the extreme political sensitivity which had been raised by a refusal to agree to

Mexican price demands in late 1978 led Deputy Secretary of State Warren Christopher (with the assistance of the Department of Energy) to negotiate directly with the Mexicans in 1979 on the price of a much reduced quantity of gas. The fact that the senior official in the negotiations was responsible for foreign policy indicates the importance of natural gas trade in US–Mexican relations at that time.

In the case of Canadian gas trade, the exporting government determined the export price (known as the 'uniform border price') as well as licensing export volumes from individual fields, for ten years leading up to 1984. With the need to gain regulatory approval for any gas imports into the USA, this could be seen as government to government negotiation of prices. While such an interpretation would not be strictly accurate, since the Duncan–Lalonde formula was not an agreement in the accepted sense of that term, it would be true to say that the US and Canadian regulatory bodies controlled the trade, although they did not negotiate directly with each other.

(*d*) *Limiting a supplier's market share: the issue of security of supply*. This type of intervention arose out of the transatlantic agreement that followed the Soviet 'pipeline' episode of 1980–82. It will be recalled that the Reagan Administration strongly objected to the expansion of West European imports of Soviet natural gas. As a result of these objections, the Administration imposed sanctions on energy-related equipment sales to the USSR by US companies after the declaration of martial law in Poland, and subsequently extended these measures to include European subsidiaries of those companies. The resulting transatlantic tensions created by US actions led to an agreement, within the framework of the IEA, to study the security aspects of West European gas supplies. The outcome of these deliberations, issued at the 1983 ministerial meeting, was that importing governments agreed to 'avoid undue dependence upon any one source of gas imports', and obtain future gas supplies from 'secure sources' with emphasis on indigenous OECD sources.

The governments let it be known that this meant Soviet gas should constitute no more than 30–35 per cent of total gas supplies in any major West European country (West Germany,

France and Italy). The legacy of the pipeline episode has therefore been to impose a limitation on European gas utilities in the percentage of Soviet gas supplies for which they may contract without incurring the attention of their governments, which will be subject to pressure from the USA to limit their purchases from this source.

This was perhaps the first time that the security aspect of European gas trade was brought to prominence. However, rather than inspiring a discussion of the real security issues, the Soviet pipeline episode illustrated the differences in governments' definitions of security of supply, compared with how commercial entities view the problem. When utilities consider security of supply, their primary concerns are domestic and technical, focusing on adequate pipeline logistics and compression. From an international perspective the primary concern is commercial reliability – both from a technical and a contractual perspective. There needs to be confidence that the exporter is both able to operate the export facilities and prepared to adhere to the terms of the contract; any interruption for either technical or contractual reasons being extremely costly for both parties.

For all of these reasons it is unwise to become overly dependent on one single source of supply – although this is simply common sense and applies to imports of any commodity. Any country, however friendly and politically stable, can develop unforeseen problems in its export trade – an obvious example being the strikes (and threats of strikes) on the Norwegian North Sea platforms which have, in 1986, severed the flow of gas to European importers. In addition, also in 1986, technical difficulties in Norwegian offshore production have caused deliveries to Continental Europe to be cut back, perhaps for a protracted period.

For much of the 1980s, Japan will remain dependent upon Indonesian LNG for more than one-half of its total supplies. Some regions outside Tokyo are entirely dependent upon Indonesia for their natural gas supplies. This is not necessarily ill-advised and has certainly caused no problems for LNG importers over the past decade; nor has it caused any great concern within the IEA. From a government perspective, therefore, natural gas security of supply issues focus on the USSR.

(*e*) *Equipment contracts*. While the four categories cited above are the most clear-cut examples of intervention, there are some less visible forms of government activity of which the commonest are sales of equipment and technology. Natural gas trade often gives rise to sizeable business opportunities since pipeline or LNG installations usually run into some billions of dollars and the contracts for construction of even part of the system are won only under conditions of fierce competition. In a world where steel and engineering industries are suffering hardship and the unemployment in these sectors is crippling individual towns and indeed entire regions, governments are naturally concerned to make every effort to bring the business to their own country. This was seen most clearly in the contracts for equipment and technology for the Soviet pipeline, where steel and engineering companies, often armed with government credits allowing attractive interest rates, strove to outbid each other in order to obtain contracts without which some might have been forced out of business.

Conclusions

There is no country in which government does not maintain total control of imports and exports of gas. Gas trade projects are usually of such a large size that governments have a habit of regarding them as crucial to the national interest, both domestically and in terms of foreign policy. Thus it would be surprising if, for example, the transition of the British Gas Corporation from the public to the private sector was to herald any significant change in the control over trade exercised by the British Government.

When looking at instances when interventionist policies have been pursued, these seem slightly more likely to occur during periods of tight gas and energy supply. For example, in the USA, the drive towards abolition of natural gas regulations governing both domestic and international developments has been accelerated by the persistence of a gas glut or 'bubble'. It remains to be seen whether a tightening of supply will be accompanied once again by imposition of government controls. However, government intervention is not necessarily directly related to market conditions (see especially types (*b*) and (*d*) above).

An interesting feature of the types of government intervention discussed here is that they do not appear to depend on the ownership structure of the utility. The exception would be in the second type of intervention, where it would be extremely difficult for a government to force a privately owned utility to sign a contract with a foreign exporter. Nevertheless, privately owned utilities like to remain on good terms with their governments in order to avoid sanctions being imposed upon them through the regulatory system, or to claim some quid pro quo in the bargaining over rate levels and structures. In the main, governments always retain the power to veto a project, but only limited ability to coerce a utility.

Commercial interests – whether companies or utilities, state or privately owned –invariably regard government intervention as detrimental to the business. However, although in many cases such intervention *has* had disastrous consequences, in some instances trade could not have taken place without government action. The action of European governments to protect the Soviet gas contracts would be one example from pipeline trade. Virtually none of the Japanese LNG import contracts would have been achieved without the encouragement and financial support of the importing government. In a few rare cases, government has shown itself wiser than the private sector; witness the refusal of the US regulators to allow large-scale LNG imports from Algeria in the late 1970s which saved a number of companies many billions of dollars of wasted investments.

The gut reaction of commercial interests to government intervention is always that the business would be much better off if left to itself. Even if true, this is irrelevant. Governments are here to stay in international gas trade: the only thing to do is to try to work with them. Those who try to work against them will invariably lose, even if time proves the instincts of politicians false.

10 NATURAL GAS DEVELOPMENT: PRICING, POLITICS AND PROFITABILITY*

Aman R. Khan

Introduction

Natural gas is gradually asserting its rightful role as an important energy source in the world. 1984 saw a dramatic increase in reserves, commercial production and use at a time when the other principal energy source (oil) was facing serious problems. At the end of 1984, proven natural gas reserves amounted to over 3.4 quadrillion scf or more than 586 billion boe, which amounts to 84 per cent of the world's known oil reserves. Compared with that of oil, the increase in marketed production was even more dramatic. As oil lost ground in the market-place, natural gas made inroads, and consumption of natural gas amounted to about 54 per cent of total world oil production in 1984. To cite a few statistics, gross production of natural gas totalled 33 mboe/d, marketed production was approximately 29 mboe/d and the international trade in gas via pipeline or as LNG was 10 per cent of gross production. Although the international trade in gas is not significant compared with that of oil, local and regional consumption now accounts for more than a quarter of global energy requirements. However, the further advance and development of natural gas on the world energy scene will necessarily be affected by structural, regulatory, economic and political factors.

After the oil price shocks of the 1970s, natural gas, which had been relegated to the position of a secondary fuel, suddenly assumed greater importance. Governments have now recognised the real value of natural gas and are keen on exploiting it as a means of displacing costlier oil imports. Moreover,

* Paper presented at the Oxford Energy Seminar, 1984. Data updated 1986.

the major oil companies, which had largely ignored natural gas as a revenue earner, moved rapidly into this area when faced with the erosion of their share of the oil business in favour of that of the host countries.

The recent decline in oil prices, increasing interest rates, and the general economic malaise which has afflicted the industrialized nations may seriously affect and stifle further growth of the natural gas industry unless proper action is taken. The vast natural gas resources cannot be exploited and developed unless sufficient economic incentives are provided and regulatory limitations are relaxed. The primary issues of pricing, politics and profitability are deeply intertwined and will significantly affect natural gas development, whether it be for use within the producing country or for international trade.

Natural Gas Pricing

With the emergence of natural gas as a major energy source for both industrialized and developing countries, its price in the market-place has become a central issue. Currently, there are three basic approaches to gas pricing:

(a) Cost-based pricing
(b) Value-based pricing
(c) Opportunity cost pricing.

Cost-based pricing methodologies evolved from the cost-of-service techniques used to determine the cost of transporting gas via pipelines. The cost calculations are based on the investment in facilities, operating costs and a predetermined, reasonable return on investment.

The most difficult problem in setting a cost-based price is to establish a rate of return on investment that will encourage the operator to develop the resource while at the same time ensuring that the delivered cost is not excessive to the buyer and is competitive with other fuels. In the past, producers have sought to maximize their return by allocating a portion of oil-related production costs to natural gas production in the case of associated gas, and by asking for the same rate of return for production of known gas resources as they receive for high-risk oil programmes. Since few concessions or production-sharing agreements provide for the type of record keeping needed to allocate costs precisely, the cost allocation finally becomes a matter of negotiation. It is arguable that most, if not all, exploration is directed towards finding oil and that the 'exploration risks' for natural gas that is discovered while looking for oil are zero. This line of argument would maintain that the rate of return for a natural gas project should be related to the risks of development, which are much lower than the risks of exploration. Host countries use this argument to support a lower rate of return for natural gas development projects. It is true that most multinational oil and gas producers have opportunities for investments that will pay a higher rate of return than gas development projects, although these may be limited.

One of the strongest weapons the host country has to assure investment in natural gas production is the link between gas production and oil production. In effect, many host countries negotiate a 'package deal' in which the producers are assured they will be able to produce the oil on their concession if they produce natural gas at a reasonable price. On the other hand, if non-associated gas is produced, the host country is in a difficult position and has to offer greater incentives to ensure development.

Value-based pricing is a sound economic and marketing concept and relates the price of natural gas at the burner-tip to the cost of the next best alternative. Definition of the next best alternative is subject to varying opinions. The seller will argue that natural gas, which is a premium fuel, competes with distillates and other light fractions whereas the buyer (especially a power utility) will opt for heavy fuel oil or coal as the next best alternative. There is the additional problem of indexing to maintain the level of market prices as energy prices change.

Even if a value-based price can be established in the marketplace, how is this price to be translated into a well-head price? If an average treatment and transportation cost is used to calculate the well-head netback price, some producers whose fields are close to the markets will reap greater profits while others with production more distant may not receive a price that would encourage development. If each supplier's netback is computed separately, there will be a multiplicity of well-head prices.

As natural gas is still considered a replacement for oil in many markets, both crude oil and a broad variety of oil-derived products have been suggested as the market proxy. Although there is little direct competition between crude oil and natural gas in the market-place, world crude oil prices are easily identified and convertible to gas prices on the basis of energy content. In the case of oil products, selection of a market proxy acceptable to both buyer and seller can be difficult. Use of a heavier oil product would favour the buyer because of the relatively lower price, whereas a light oil product would favour the seller by maximizing the well-head price. The current trend is to use a combination relating to the market price of distillate fuel oil and furnace, or residual fuel oil, as the parity pricing base. The weighting factors for light and heavy oil products may be arbitrary but should be related to the fuel market profile.

Opportunity cost pricing is based on establishing the highest market value for the gas, assuming a number of buyers. The opportunity for large volume sales would usually be the export market for pipeline gas, LNG or for converted products such as methanol, fertilizers or petrochemicals. The opportunity cost is determined by the market value of LNG or converted product and then calculating the netback value of gas at the well-head, that is, the export price minus the conversion, transportation and treating costs. The opportunity cost is subject to many uncertainties. Since there is no standard or marker world price for natural gas, each project has to be evaluated on the basis of its own costs and the competitive market.

Sellers who use projected LNG prices as the opportunity cost may find that the projections are not realized and the price in the future will be considerably below estimates. If the gas is to be sold in the form of chemicals, future competition may force world prices below present levels and the netback could become negative.

The risks inherent with this type of pricing are only too obvious in this period of depressed demand and falling oil prices. Because export projects are highly capital-intensive, the seller must negotiate long-term sales contracts for the products, whether they are natural gas or gas-derived chemicals. If the price of the base fuel against which the contract prices are indexed falls, project revenues will drop and costs may not be covered. In point of fact, this is exactly what has happened in the past two years as prices have fallen because of increased competition and decreased demand. Therefore, opportunity cost-based prices should be viewed with a great deal of scepticism and used only as approximate guidelines.

Project Profitability

The development and utilisation of natural gas in any project today requires a large investment, and profitability has an important bearing on the project's success and future expansion. Typically, a grass roots gas development programme requires investment in exploration, appraisal and development drilling and construction of the field and delivery facilities. These will consist of gathering systems, separation and purification plant, a pipeline transmission system and a gas distribution network. Other downstream facilities may include an LNG export plant or plants to convert natural gas to chemicals or other products. In the initial stages of a project, large investments are made and costs are incurred in the exploration, development and construction phases resulting in a negative cash flow. Thus, production of the first volumes of gas requires large initial expenditures but additional volumes require lesser cash inputs until the system is at maximum capacity. For a desired rate of return over the project, the average price of gas that will generate the desired rate of return over the life of the project can then be calculated.

If all project components are under one management, such as a government or a national oil company, the profitability standards may be somewhat different than those applying if the various elements of the project involved outside participation such as that of a producing company.

With the large investments required for exploration and discovery, it is almost inevitable that outside participation is necessary, especially in developing countries. This raises the question of the arrangements for sharing the potential profits resulting from successful exploration and development with the outside investors. While some countries still use the traditional licensing and royalty arrangement, others have profit-sharing agreements. The latter are gaining more prominence in the lesser-developed countries where natural gas is just beginning to enter the market.

The mechanism for determining the government's share of profits varies and may be based on the relationship of project profitability with royalty rates, profit, taxes, etc. If the government share, including taxes, royalties, production shares, etc., is high and inflexible, exploration programmes will be curtailed and hydrocarbon development restricted. Conversely, if the government share is not sufficient, adjustments on taxes, royalty rates, etc. will be required at a later stage of the project.

In a recent article, the World Bank concluded that rate-of-return (ROR) based profit sharing meets the objectives of both government and investor better than more familiar profit-sharing alternatives.[1] According to this concept, the government share of profits is flexible and can be adjusted as a function of the outside investor's discounted cash-flow rate of return. The government share under ROR-based profit-sharing is therefore a function of achieved, rather than estimated or anticipated, profitability. It is claimed that the ROR system is responsive to the major determinants of project profitability such as price changes, reserve life index, operating costs, timing of cash flows and cost of capital. Other profit-sharing alternatives, where the government share is tied to production, revenue generated, cost recovery or price, have the disadvantage that they do not respond to *all* of the determinants cited above.

In a typical ROR-based profit-sharing system, the ROR will be negotiated and agreed upon (perhaps in the range of 15–20 per cent) prior to incurring expenditures in exploration and production. The government share may be in the 45–55 per cent range. Until the contractor has recovered his initial investment and the agreed upon ROR, the government share does not increase. However, once this threshold has been reached, the government share can escalate by incremental steps.

[1] McPherson, Charles P. and Palmer, Keith: 'New approaches to profit sharing in developing countries' in *Oil & Gas Journal*, Vol. 82/No. 26, June 25, 1984.

Currently, at least a dozen developing countries have adopted the ROR-based profit-sharing system.

The standards of profitability will vary depending on the viewpoints and perspectives of the parties involved. In the case of a gas-exporting country, the government will look at the impact on the national economy. Apart from the revenues from royalties and taxes, etc., the project will have an impact on the country's balance of payments, its borrowing capacity and the long-term commitments it must make. If the export project is managed and operated by a national company, the latter would have to operate it profitably while recognising the constraints imposed by the government, which owns the company. Yet another viewpoint is that of the international oil company which holds a working interest. Its principal interest is in the profitability of the venture with little or no concern for national economies at either end of the export–import chain. Finally, there is the importing country which has to consider the impact on its economy and balance of payments as well as the effects of reliance on energy imports of a long-term nature.

Issues in Natural Gas Pricing

Based on the profitability of a project, the delivered cost of natural gas can be estimated. The gas price must meet all fixed and variable costs related to field development, production and system operation and generate a reasonable return on investment. As the system expands, each increment of gas production has to be priced at least at a level sufficient to meet the marginal costs of that increment. In this respect, it is better to consider the long-term marginal costs, namely the change in total costs over the whole production period as a result of the increment in gas supply. Of course, there are risks in setting a long-term marginal price on the basis of projected development programmes and sales. The market may develop more slowly than anticipated, in which case the long-run marginal cost will have been too low, or reserves may last longer than expected and the marginal cost will have been too high. Clearly, this is an important aspect to be taken into account when setting gas prices.

Obviously, national objectives must be met and they have a

bearing on gas pricing. The gas price should be set at a level that will encourage industrial growth. This does not mean that the gas should be sold very cheaply but, rather, that gas produced and sold locally should be priced such that it competes with other fuels and is substituted for any imported oil products. In developing countries where thermal power plants use fuel oil, gas prices can be set at a level to minimize operating costs and therefore the cost of electric power to the consumer. However, this does not necessarily imply that the government should subsidise the gas price for internal usage unless the benefits to the national economy present strong justification.

One factor which has been largely ignored in the past is the non-renewable nature of the resource base. Natural gas, like oil, is an exhaustible resource and it is imperative that as development and exploitation of a gas field take place, investments are devoted to continued exploration for finding and developing new fields. The money for financing these efforts should be generated from existing revenues and therefore gas prices should be structured to generate the necessary capital.

There are varying opinions on the subject of pricing an exhaustible resource, the so-called depletion value. One alternative is to produce the gas and sell it at a price competitive with oil in the present day market, say at $2.50–3.50/mcf if oil prices are in the range of $15–21 per barrel. On the other hand, the gas field could be shut in and produced much later, if and when oil prices escalate, thereby increasing the value of the gas. The final decision on which path to follow depends not only on the revenues generated but also on national policies and the magnitude of the non-renewable energy resources. In any case, it would be prudent and highly desirable that the price of gas be such that it provides for funds to be set aside for exploration and production of replacement gas reserves as well as for the development of other alternative energy resources.

All of these issues in gas pricing lead to a central question: what should be the degree of government regulation and control? Although this may vary from country to country, there are examples that provide guidance on how *not* to set gas prices.

Natural Gas Price Regulation

A classic example of the ill effects of government regulation of natural gas prices is that of the USA where the well-head price of gas was regulated much too tightly during the 1950s and 1960s. During this period, the regulatory bureaucracy held well-head prices down to 10–15 cents/mBtu. The low prices caused demand to exceed supply and resulted in a gas shortage. Moreover, the low price provided no incentive whatsoever to producers to make the investments required for new gas production and expansion of capacity to meet the growing demand. The gas shortage of the 1970s resulted in legislation to decontrol and deregulate prices and the situation was reversed with the increase in gas prices. However, the price advantage that natural gas held over competing fuels for many years has gradually disappeared and presently there is a surplus of natural gas supplies in the USA.

An example of a developing country which had a similar experience is Pakistan. The Sui gas field which was discovered in 1954 is the mainstay of the country's gas supply. Until recently, no incentive was offered to foreign companies to invest in exploration or to develop the existing potential. The well-head price of gas prior to January 1982 was 6 cents/mcf, one of the lowest in the world. Demand for gas increased in the 1970s as the prices of oil and oil products went up and by 1980 the country was faced with a severe gas shortage. The result was a curtailment of gas to industry which affected production and the economy. As a result, the Government took steps to increase the price of 'old gas' by stages. In 1983 the well-head price of Sui gas was increased to 35 cents/mcf. Furthermore, a new formula was introduced in 1983 for calculating the producer's price for new gas. This formula is as follows:

> To improve the incentive for production of natural gas from new gas fields the Government of Pakistan is prepared to negotiate with the prospective producers a base price for new gas which would ensure achievement of an internal rate of return of 12% on the basis of discounted cash flow. The cash flow would include net receipts after taxes and all expenditure on the project both development and operating inclusive of taxes and duties.

> The period of recoupment of capital investment will be twelve years or the estimated life of the field whichever is less. Ordinarily the cash flow will be projected for a period of 12 years after commencement of commercial production. The base price so determined will be indexed to international price of light Arabian Marker crude posted by OPEC.

In 1984, the well-head price was increased once again to 62 cents/mcf, and in mid-1985 the whole policy was further changed. According to informed sources, it appears that the pricing policy, at least until 1985, was not attractive enough to entice investors.

To sum up, the two examples cited above indicate what can happen to the development of gas if over-zealous regulation is allowed to dictate the price of gas without an understanding of the economic forces at work in the market-place.

Pricing and Politics

The price of natural gas to the consumer in the developed countries is determined largely by inter-fuel competition and to a lesser extent by national policy and regulation. In the developing countries, this is tempered sometimes by political considerations which may have little or no relation to economic factors. For instance, there may be a two-tier pricing mechanism: one price for export as LNG pipeline gas or gas-derived products, and a lower price for the domestic consumer. The latter is then partially subsidised by the export earnings. Sometimes the reverse occurs when the export industries receive gas at a lower price and the domestic consumers pay higher gas costs.

In a much broader context, political considerations have a significant influence on the gas trade whether it be from the corporate, national or institutional viewpoints. Historically, the major oil companies always viewed exploitation of gas in their overseas holdings as less profitable than oil because of the absence of a well-defined international market for gas and because of its low price. However, with the growth of the market and the increase in prices, the majors are becoming more involved in the gas trade. Nevertheless, the politics of oil companies militate against the displacement of their prime product (oil) by gas. The international gas trade requires long-term contracts with no room for spot markets. Thus, the trade is not as attractive to those oil companies that seek a fast and high return on investment similar to that which they obtain in the exploration for and development of oil. Furthermore, if there are no potential export markets for gas discovered by a foreign oil company in a country, and if the national interest requires that the gas be used to meet internal demand, there is always the danger that the price set by the government will not be sufficient to meet the company's minimum requirements for return on investment. Thus, corporate politics will inevitably be dictated by corporate profits although in some cases, such as the utilisation of associated gas, foreign companies may be forced to compromise their position in order to maintain their profits from oil production.

At the national level, there is considerable political pressure to maximize earnings on gas exports. This has been most evident in the past six or seven years in the LNG trade. Figure 10.1 shows how LNG prices increased in the late 1970s as oil prices increased. Then, as oil prices dropped in 1981, LNG prices fell parallel with this decline. The concept of linking LNG prices with competing fuels such as crude oil or fuel oil was propounded by sellers and accepted by buyers, although there was a difference of opinion on whether the pricing parity should be at the exporting or the importing terminal. In the case of the LNG export trade to Japan, both buyer and seller have accepted the principle that price parity will be c.i.f. the port of entry.

However, in the case of Algerian exports of LNG to Europe and the USA, Algeria has maintained that the price parity should be linked with the export price of crude oil. The formula proposed by Algeria is as follows:

$$P_g = P_{gi} + K(P_c - P_{ci})$$

where

P_g = Current f.o.b. price of natural gas per mBtu
P_{gi} = Initial contract price of natural gas per mBtu
P_c = Current f.o.b. price of basket of crude oils
P_{ci} = Initial price of basket of crude oils
K = Factor to be negotiated.

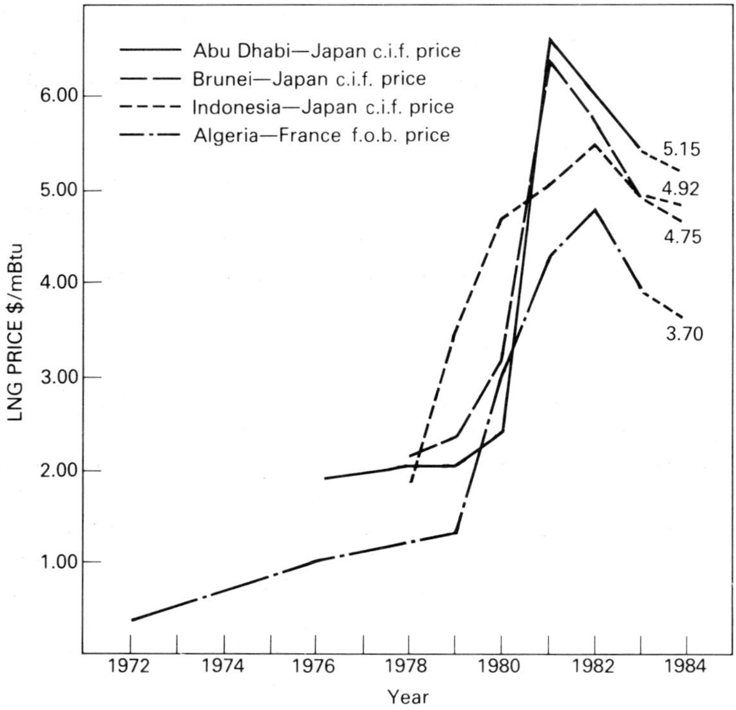

Figure 10.1 Selected LNG Prices. 1972–84.

The Algerians have used this formula in negotiations or renegotiations on LNG contracts with Belgium, France and Italy. Distrigaz of Belgium was the first buyer to accept this formula in January 1981 and accepted an LNG f.o.b. base price of $4.80/mBtu linked to the price of a basket of crude oils. The following year, the French agreed with Algeria to an LNG f.o.b. base price of $5.12/mBtu indexed to a basket of eight crudes. But in this case, Gaz de France resisted the price increase demanded by Sonatrach. The French Government agreed to subsidise the difference between the two national bodies, which amounted in fact to a 'political' payment.

A similar deal was made by the Algerians with Italy for pipeline exports of gas ($4.41/mBtu f.o.b.) with the Italian Government agreeing to pay part of the price as a subsidy. These are clear examples of political pressures being applied in the free trade process. However, other buyers of Algerian gas such as Spain and the USA have not yet agreed to the concept, nor have other gas exporters such as Abu Dhabi and Indonesia adopted the principle of pricing parity f.o.b. rather than c.i.f. Common sense dictates that in international trade gas has to be competitive in the market-place and that pricing gas at parity with crude oil or oil products at the point of export will certainly not permit this.

In 1980, OPEC endorsed the principle of oil–gas price equivalency and set up a committee to study the issue and develop some uniformity among member countries on the question of export gas prices. These efforts were unsuccessful for a number of reasons. Potential buyers had access to gas from non-OPEC sources such as the USSR, Canada and Norway and were able to obtain supplies at relatively low prices. Also, in the last two years there has been a surfeit of energy supplies and a decline in energy prices. Whether these efforts by OPEC will be successfully pursued in the future remains to be seen. But this is another example of political influence imposed at the institutional level on the gas trade.

Finally, one must mention political motives that are evident in the international gas trade as an effort to forge stronger trade links between countries. In the late 1960s, the USSR started importing natural gas from Iran and Afghanistan. In both cases, the USSR ensured that other trade ties with these countries were expanded and, in the case of Afghanistan, so was their political influence. Soviet economic and military aid to Afghanistan was increased and paid for by the Afghans with the exported gas. This bartering of gas was also practised in the Soviet–West European gas supply contract whereby the USSR partially traded its gas for equipment manufactured by West European suppliers. Naturally, this increases the value of natural gas to the buyer's economy.

Since the supply contract for Soviet gas to Europe is long term, this also creates a dependence on energy supplies which from a geopolitical viewpoint strengthens the influence of the USSR. Today, the USSR is the world's biggest producer of natural gas and has 39 per cent of the world's proven reserves. This potentially places it in a very strong position as the dominating exporter with the ability also to control prices in the international market. Recently, the USSR offered short-term gas supplies (spot deliveries) to Belgium at prices lower than those of Dutch gas. The Soviet spot price at the Belgian border was \$3.15/mBtu, compared with \$3.70/mBtu for Dutch gas and \$5.50/mBtu for regasified LNG from Algeria.

While these efforts by the USSR to penetrate the West European market may be of concern to other suppliers, such as Norway and the Netherlands, it is clearly in the interests of buyers to diversify sources of supplies. It is politically advantageous for exporting countries like Algeria and Indonesia to broaden their market and increase the number of buyers, just as it is prudent for Japan to have multiple suppliers of LNG to meet its demand.

Conclusions

The natural gas market today is faced with increasing inter-fuel competition and the higher costs associated with finding and developing additional reserves. World-wide reserves and production continue to increase as gas is developed and utilised by countries seeking to decrease costlier oil imports. Traditionally, gas prices have followed oil prices and have usually been lower. On the other hand, there has been no unified pricing policy for gas in the international market. Border prices vary from \$2.00/mBtu to \$5.00/mBtu in different parts of the world. Domestic prices range from 10 cents/mBtu in the Middle East to \$7.00/mBtu in the USA. However, there are certain basic requirements and objectives to be considered in gas pricing.

First and foremost is that the standards of profitability should ensure a reasonable return on investment which will encourage companies to explore for and develop gas to replace that which is consumed. In developing countries the price to domestic consumers should reflect the true costs of production and not be subsidised. It should be designed to encourage industrial growth and substitution for imported oil products.

Another minimum criterion in gas pricing is that of the long-run marginal cost, which takes into account future system expansions. Where gas is traded, the border price should be competitive with other fuels and pricing parity should be in the market-place, not in the country of origin. Also, stability of supply is an essential component that cannot be forgotten in the international trade. Natural gas is a depletable resource with an intrinsic value, which should be reflected in the well-head price.

Finally, in order to promote competition in the energy market, governments should refrain from enacting regulations on the well-head pricing and use of gas. The removal of constraints on the exploration and development of natural gas and the encouragement of strategies for its growth in the energy markets will go a long way towards ensuring future energy supplies at minimum costs to the consuming public.